DEEPWATER DECEPTION

DEEPWATER DECEPTION

The Truth about the Tragic Blowout and Perversion of American Justice

ROBERT KALUZA

and

MARYANN KARINCH

DEEPWATER DECEPTION

World Ahead Press is a division of WND Books. The views and opinions
expressed in this book are those of the author and do not necessarily reflect
the official policy or position of WND Books.

Paperback ISBN: 978-1-946918-16-1
eBook ISBN: 978-1-946918-17-8

Printed in the United States of America

CONTENTS

DEDICATION

This book is dedicated to those who died in the *Deepwater Horizon* blowout and to their families:

Jason Anderson, 35, Midfield, Texas
Aaron Dale Burkeen, 37, Philadelphia, Mississippi
Donald "Duck" Clark, 49, Newellton, Louisiana
Stephen Ray Curtis, 40, Georgetown, Louisiana
Gordon Jones, 28, Baton Rouge, Louisiana
Roy Wyatt Kemp, 27, Jonesville, Louisiana
Karl Kleppinger Jr., 38, Natchez, Mississippi
Keith Blair Manuel, 56, Gonzalez, Louisiana
Dewey Revette, 48, of State Line, Mississippi
Shane Roshto, 22, Liberty, Mississippi
Adam Weise, 24, Yorktown, Texas

To this list, we wish to add Don Vidrine, Well Site Leader on *Deepwater Horizon* who was unjustly indicted after the blowout and subsequently passed away due to illness.

.

ACKNOWLEDGEMENTS

On February 25, 2016 at approximately 6:30 p.m., a twelve-person jury walked back into the US Federal 5th Circuit Courtroom in New Orleans after two hours of deliberation with a "not guilty" verdict, ending my unbelievable and unexpected American injustice. For close to six years I observed how politics supersedes justice in America and how a wealthy powerful corporation can legally scapegoat its own employees to hide real culprits and hide the truth. After the swift verdict, many on our legal team asked what I would do next; my immediate reply was that I had to write a book about this not too uncommon injustice because too many Americans have a false trust that the American justice system is consistently moral and ethical. It took two years of very hard work to bring this book to the public.

I wish to whole-heartily thank the entire legal team at law firms Smyser Kaplan & Veselka and Quinn Emanuel Urquhart & Sullivan for a thorough, compelling and never-in-doubt successful defense. I especially want to thank the core attorney team of

Shaun Clarke, David Gerger, Dane Ball and David Isaak. The entire legal team did an outstanding job! I have to thank two former fellow BP colleagues Jim Smith and Eddie Coleman; Jim for checking and confirming my engineering calculations and Eddie for volunteering to help in every way possible. Thank you.

And finally, I must acutely thank Maryann Karinch, my book agent and co-author. Without her great assistance and guidance this book may not have ever been published. Maryann adapted my engineering report writing style into a book that can be read and understood by ordinary people. I think she did a remarkable job.

—Bob Kaluza

Thank you to Bob Kaluza for inviting me to partner with him, for arousing the storyteller in me, and for piquing my curiosity about the complicated science related to offshore deepwater drilling. Bob is a man of remarkable intellect, integrity, and conviction and it was a pleasure to plunge into this publishing adventure with him.

My thanks also to my partner, Jim McCormick, whose engineering background and continued interest in the quality of my work were invaluable in this effort. He had to explain a lot of engineering concepts to me and that often required both patience and diligence. I also have to credit him with naming the book—he's a genius at capturing powerful ideas in just a couple of words.

I am very grateful to Shaun Clarke and Dane Ball for their review of the manuscript and additions of valuable insights. I am similarly grateful to Judith Bailey for her superb copy editing. Judith—you are my friend and one of my favorite brilliant people.

Thanks also to my many friends and colleagues who kept asking, "How's that *Deepwater Horizon* book coming and when can I read it?"

—*Maryann Karinch*

INTRODUCTION:
FACT AND FICTION

You sit down to watch a movie about a disaster— an explosion on an offshore drilling rig that killed eleven people. The director grabs your attention by flashing emails that foreshadow the tragedy. The emails introduce some of the key characters, as well as minor ones with crucial information—everyone from day-to-day operations guys to corporate officers.

This movie has not been made, but I invite you to use your imagination.

As the opening credits roll, you hear Warren Zevon's "Disorder in The House." Our movie slowly scrolls a string of emails like the following, in reverse chronological order, preceding a tragedy that occurred on April 20, 2010.

From: Daigle, Keith G
Sent: Wed Oct 07 11:42:45 2009
To: Guide, John
Subject: RE: Deepwater Horizon Rig Audit
Importance: Normal

. . . Everybody wants to take a shot a[t] the leader at one time or another eh!!!

Best regards,
Keith Daigle
Operations advisor E&A
WL-4 379A
"Communications Should Never Be the Weakest Link"

From: Guide, John
Sent: Wednesday, October 07, 2009 6:40 AM
To: Cocales, Brett W; Daigle, Keith G
Subject: FW: Deepwater Horizon Rig Audit

See below, Norman throwing DW Horizon under the bus.

From: Little, Ian
Sent: Tuesday, October 6, 2009 5:17 PM
To: Guide, John; Daigle, Keith G
Subject: FW: Deepwater Horizon Rig Audit

John,

fyi – I assume these are the same issues that are on the audit list – there are some that were not on Kevan's list, but that is ok. I have talked with Harry and told him we are working with Transocean to close out the actions. We do need to think about how and who can help us to verify that these are closed out. Let's discuss when we go through the list.

Thanks, Ian

From: Thierens, Harry H
Sent: Tuesday, October 6, 2009 10:10 AM
To: Little, Ian
Subject: Fw: Deepwater Horizon Rig Audit

Sent from my BlackBerry Wireless Handheld

From: Wong, Norman (SUN)
To: Thierens, Harry H; Sprague, Jonathan D
Sent: Tue Oct 06 15:52:08 2009
Subject: Deepwater Horizon Rig Audit

Harry John

As Wells Director and Wells Engineering Authority, I just wanted to bring to your attention the most significant findings from the recent rig audit of the Deepwater Horizon. You may have been made aware of this already but just in case you were not, thought it best you should know. Although I did not perform the audit I am happy to discuss this matter with you.

- Closing out of the last audit recommendations had no apparent verification by BP. Consequently a number of the recommendations that Transocean had indicated as closed out had either deteriorated again or not been suitably addressed in the first instance.
- Control of work issues identified specifically with isolation permit process and integrity of mechanical isolations
- Overdue planned maintenance considered excessive 390 jobs amounting to 3545 man hours. With the recent move from Empac to RMS II maintenance systems, and revised maintenance, scheduling the back log does not look as though it will improve
- The iron roughneck could not be made to operate from Cyberbase unless the anti collision system was in override

- Top drive guard is not fitted with a safety sling, not only is this an NOV requirement but also a lesson learned from industry incidents, including one on this rig, where the guard had been knocked off due to equipment clash.
- Annual drawworks maintenance routine overdue since February 2009, includes critical checks on the braking system
- Test, middle and upper BOP ram bonnets are original and out with OEM and API five year recommended recertification period.
- The port aft quadrant watertight dampers failed to close when tested
- The starboard aft quadrant bilge and ballast valves, ballast pump and tank sounding system where [sic] rendered inoperable due to a process station (PCU 18) card failure
- Three out of four electric bilge pumps were tested, all three failed to achieve suction due to defective priming systems.
- Emergency bilge suction check valve integrity checks concluded valves were passing
- Several hydraulic watertight door issues concerning both operability and functionality. Insufficient onboard spares to make repairs
- Just one of the eight seawater cooling pumps was totally defect free. Twp of the defective pumps were identified during the previous

audit (January 2008) while some of the defective pumps could be operated, four pumps were deemed non operational.

Regards
Norman Wong
Tel +44 (0) 1932 739526
Mobile +44 (0) 7780 956533
BP Exploration Operating Company Limited, Registered office: Chertsey Road, Sunbury on Thames, Middlesex, TW167BP.
Registered in England and Wales, number 305943

These are authentic emails, included as Defendant's Exhibit 95 in my February 2016 trial. They raise many questions that propel a story of high-stakes intrigue, tragic loss of life, and closed-door deals between a corporate behemoth and the United States Department of Justice. These emails are like a flashlight in the dark: They shed light on bits and pieces of decision-making driven by BP's desire to save somewhere between $15 million and $90 million by delaying maintenance on an over-burdened rig.

Someone—some person or persons—had to be blamed for the spectacular disaster that caused the deaths of eleven people on the *Deepwater Horizon* rig. People who are angry and hurt at the loss of loved ones and severe environmental damage aren't satisfied just having a company be the villain, and the Department of Justice took that to heart.

BP and DOJ gave the public what it wanted by offering me up, along with my colleague, the late Don Vidrine. After nearly six years, four of which involved restrictions on my liberty (like not being able to answer my phone), my case went to trial.

The jury that looked at all the evidence related to those deaths and to the supposed violation of the Clean Water Act took less than two hours to proclaim my innocence. Here's why: The real story is so compelling and airtight, that no thinking human being would ever conclude that I had contributed to the deaths of those people and the ecological catastrophe associated with them.

What I tell you in this book is that true, compelling story as well as the airtight evidence related to the real cause of the tragic blowout and perversion of American justice.

The *Deepwater Horizon* disaster is a spectacular foundation for strong storytelling and a fictionalized version of what happened. In terms of media, that's where we live right now: Make a good story better by tweaking it and more audiences will pay to see it.

I have no problem with using this story for entertainment, as long as entertainment is not confused with truth. Matt Brennan, *LA Weekly* film critic, said this when commenting on the movie *Deepwater Horizon*: "Any film that takes a real life subject and dramatizes it is automatically opening itself up for scrutiny."[1]

Each of the 115 people who lived through the blowout on *Deepwater Horizon* has a story to tell. Each has a point of view about the events as well as expertise in some aspect of rig operations and drilling that gives that person a unique slant on the horrific events that killed eleven of our co-workers. In short, there is only one set of facts of the *Deepwater* disaster—which is what I've aimed to give you in this book—but there are probably 115 versions of "the truth." Another way of saying this is in the book *Nothing But the Truth*: "What we perceive as truth brings at least three interrelated elements into play: imagination, belief, and experience."[2]

In creating a feature film based on true events, filmmakers may hear from real people in the story who have different memories of what happened. And because it's not intended to be a documentary, the movie also weaves in the imagination, beliefs, and experiences of writers, actors, and the director. In other words, lots of different versions of "the truth" come into play. There is nothing to criticize in that. But people who watch *The Wizard of Oz* know it can't possibly be true, whereas people who watch Oliver Stone's *JFK* could easily come away convinced that there was a conspiracy to kill President John F. Kennedy. Similarly, moviegoers who see a film like *Deepwater Horizon* are naturally going to be inclined to accept it as a true, or mostly true, representation of what happened. And even if they assume that not every part of it is accurate, how are they going to know what to accept mentally as fact and what to recognize as conjecture—the product of dramatic license?

To compound the confusion over fact versus fiction, actors may portray characters based on real people, but they are not trying to imitate them in a robotic fact-based way. They are trying to capture them authentically, and that authenticity lies in their ability to portray the person in the context of the scripted drama. They are using the lines created for the movie to infuse a character with life. John Malkovich is a fine actor, who portrayed a character named Don Vidrine, but he shouldn't be mistaken for the actual BP rig supervisor Don Vidrine. The two Dons are very different.

I know there is a high likelihood that you are reading this book after exposure to the movie, to *60 Minutes* coverage on television, and to lots of articles and news clips. That's why I wanted to distinguish upfront between elements that constitute an engaging story and facts that give a comprehensive picture of reality.

Facts are not necessarily boring, by the way. They are your basis for intelligent conversation and, in some cases, involve even more dramatic moments than those in the movie. In this case, the heroes and villains in real life are just as intriguing as those in the movie.

With this book, I invite you on an adventure, an investigation, and a manhunt. You are looking for the real killers—the people directly responsible for causing the *Deepwater Horizon* disaster. With that aim in mind, I will respectfully disagree with movie's producer, Lorenz di Bonaventura, who tried to explain why he wanted to back away from finger-pointing. In an interview with the *Los Angeles Times* he said, "In those kinds of the events, there is no black and white."[3]

I'm here to give you the evidence that there is.

CHAPTER ONE

3 MISTAKES

Who caused the death of eleven crew members aboard the *Deepwater Horizon*?

This is a people question—a "who" not a "what"—so it can't be answered with the name of a company. BP, which operated the floating rig that went up in flames on April 20, 2010, paid fines and compensation related to the deaths of those men. But BP cannot serve time in prison.

There might be an answer to "who" found in this story. Woven into it are the names of individuals who could logically be considered responsible for the tragedy. You may even have me on the list if you ask yourself the tough questions about who *might* be held accountable. After all, I was one of two BP rig supervisors indicted for the crimes of manslaughter and a violation of the Clean Water Act—although the manslaughter charges were dropped, and it took less than two hours for a jury to find me innocent of any other charge. About half way through the book, you will have a clear idea of who should be held accountable for the disaster. By the end, you will not only have a list, but you will know precisely how their actions contributed to the tragedy. And then, you will very likely demand to know what Jesse Eisinger explores in his book *The Chickenshit Club*, that is, why the Justice Department fails to prosecute executives.[1]

Questions about guilt or innocence penetrate every description of what happened on *Deepwater Horizon*. For that reason, you will be whittling down a list of about two dozen people on the rig and onshore to a much smaller pool.

The story begins with three mistakes. To give you a context for them, and for knowing why they led to the disaster, I'll start by taking you onboard *Deepwater Horizon*, the offshore drilling rig that was owned and staffed by the Swiss company Transocean Ltd. and leased to BP, the British oil and gas company where I worked as an engineer and rig supervisor.

Deepwater Horizon And The Macondo Prospect

Deepwater Horizon wasn't just built for water, it was built for ultra-deepwater. Of the different types of rigs used in the Gulf of Mexico this one was categorized as a semisubmersible —a floating rig that could move through the water like a marine vessel yet remain dynamically positioned to a spot while it was enabling drilling work on the ocean floor. It was designed to release from its position and float away, either because it was needed somewhere else, or in the event of an emergency.

Deepwater Horizon had special stability characteristics that made it possible for this rig to drill the deepest oil well in history about 250 miles from Houston. When the job was done at that site, the rig was moved through the water to explore the next site identified as a likely source of oil.

The job of *Deepwater Horizon*'s crew was exploration. They were hired to drill for oil, not to extract it; the oil and gas beneath the surface of the earth were never supposed to make their way all the way up the pipe and onto the deck of the rig.

The last location of the rig was known as the Macondo Prospect. A prospect is a rock formation that geologists have identified as being the possible location of a petroleum reservoir. Macondo is a BP company nickname for the site and it refers to an area officially known as the Mississippi Canyon Block 252 (MC252). That's how the Minerals Management Service (MMS) of the U.S. Department of the Interior (DOI) designated it in granting BP the rights to drill there in March 2009. Macondo is a name lifted from a Gabriel Garcia Márquez novel; some BP employee won an internal contest by picking it. When we talk about the Macondo well, we're talking about BP's drilling operation at the Macondo Prospect.

When you stood on the rig floor of *Deepwater Horizon*, you were on a powerful, massive vessel. At 396 feet long, it was 36 feet longer than a football field, and at 256 feet wide, it was nearly 100 feet wider. It could operate in waters up to 8,000 feet deep, so its position at Macondo was comfortable; the water depth to the well head at this site was 5,067 feet below the surface of the water—an impressive depth but well within the operating envelope of the rig.

Imagine standing on that rig floor with a vision of what connected you to the sea bed. Just below you was a heavily reinforced pipe called a riser that went all the way from the rig to equipment that sat on the ocean floor 5,000 feet beneath the water's surface. This single length of pipe connecting the massive rig with its most important safety mechanism had an inside diameter of only 19.5 inches and outside diameter of 21.5 inches, inches, making it an inch thick all around. Every day, as the federal government requires, one of *Deepwater Horizon's* two remotely operated underwater vehicles (ROV) inspected the riser. It was like letting a dog out on a leash to sniff for trouble.

The equipment that rested at 5,000 feet below was a combination of mechanisms called blowout preventers; together they are referred to as BOPs, BOP equipment, or the BOP stack. Those of us who have expertise in offshore drilling operations know how many variables there are when dealing with the power of nature, so we build rigs with a great deal of redundancy related to safety. The BOP stack is a gigantic device weighing about 700,000 pounds (350 tons) that sits on top of the well head. A typical subsea BOP stack is about 50 feet tall, 15 to 25 feet across and 7 to 10 feet thick. This one was 53 feet tall. Nearly the height of the six-story courthouse in which I would ultimately go to trial.

BOPs are often referred to as "the stack" because they are composed of a vertically stacked series of single, and sometimes double, independent sections. Each section is designed to close or open around pipe or to close off the open hole completely; each section opens and closes independently, so in combination, the series of BOP sections create multiple and different ways to cut off the flow of oil or gas out of the well if needed. If one BOP section fails to seal off the problem, another independent section can be activated to do it. And with multiple redundant sections, BOPs should never fail. Subsea BOPs are designed and built to be fail-safe.

There you are: Standing on an expanded football field, with about 5,000 feet of thickly reinforced pipe under your feet that is attached to a 350-ton stack of safety equipment seated on the ocean floor. From pontoons to the top of the derrick is 385 feet. The main deck is 136 feet above the pontoon base, so from the main deck, the top of the derrick rises above you by nearly 250 feet.

Keep in mind that the depth to which the rig would sink in the water—it's draft—depends on its load at the time. The drilling draft for *Deepwater Horizon* ranged from about 67 feet to 75 feet. From the main deck down to the water would therefore be somewhere between 61 and 69 feet. The lifeboat deck is 12 feet below the main deck. If you had to jump into the ocean to save your life, whether it was from the main deck or the lifeboat deck, you would be jumping the equivalent of six or seven stories.

Far below the rig, at a depth of 18,360 feet, is the bottom of the well you know as Macondo. A drill pipe goes from the ocean floor to the bottom of a shaft that is reinforced with cement on all sides. That shaft, called a casing run, is not as wide as you might think. At its widest, near the connection to the BOP stack, it's just three feet. At its narrowest at the bottom of the casing—18,304 feet—it's just seven inches. Below that, the open hole to the bottom of the well is just 8.5 inches wide.

There is pressure pushing upward from a reservoir like Macondo that's filled with oil and gas, but it is countered by a specially formulated drilling fluid called synthetic oil-based mud (SOBM). Whenever I refer to mud in this book, I'm talking about SOBM. Mud is circulated down the cement-encased shaft and back up the sides on an ongoing basis. It keeps the oil and gas down in the reservoir where it belongs until another type of rig comes along to liberate the oil and gas from below the ocean floor.

Mud is also denser than seawater, of course, so the pressures on the inside of the casing filled with mud are greater than those from the formation pressures, or reservoir pressures, outside. In short, as long as *Deepwater Horizon* could keep the oil and gas from pushing upward, there was a stable well.

To summarize, when you stood on the rig floor of *Deepwater Horizon*, you stood on more than 50,000 tons of metal that could hold more than two million pounds and was protected by 350 tons of safety systems seated on a cement-lined shaft filled with mud and drilling equipment that reached to a depth of 18,360 feet.

It would take a lot of power—and a lot of things going wrong—to destroy something that massive. It all boils down to three mistakes.

Three Mistakes

We now plunge into the drama of the *Deepwater Horizon* tragedy by beginning with three mistakes. They shed light on the decisions that key people made in the years, months, and days leading up to the blowout. These mistakes are not listed in order of sequence because they all occurred over time and are interrelated. They do suggest order of importance, though, with the first one being the lapse in judgment that directly caused the *Deepwater Horizon* blowout: It's the reason every critical component in the BOP stack failed.

Mistake #1: Hoping the BOP would work instead of taking action to ensure its integrity

When an oil and gas driller arrives on an offshore drilling rig deck to begin his work, he asks the person he is relieving "Any problems with the rig or the well?" The answer you want to hear at the hand-over discussion is "No," delivered without doubt or hesitation. But every experienced driller understands that well conditions can change quickly. A "no" at 7:15 could be a "yes" at 7:16. The one, consistent action that minimizes the possibility of such a sudden change is maintaining equipment on schedule and doing inspections and testing in accordance with federal regulations.

But as anyone who has worked on a rig will confirm, Mother Nature can offer huge surprises. Nature is not only composed of plants, animals, and rays of sunshine; it is also volatile gasses and potentially explosive fossil fuels. Drilling sometimes unleashes the power of nature in unexpected ways, regardless of the expertise of engineers and the drilling crew.

We know that. This is precisely why rigs involve so much redundant safety equipment. The nerdy details about this are in Appendix A (if they interest you), but right now, we are focused on nature's surprises and the processes for making sure they don't pose a threat to human life and the environment.

When we talk about geologic formations—downhole formations—in a well like BP's Macondo, we are talking about formations that occur below the sea bed. In this case, it would be formations that occur from the sea floor at about 5,000 feet below the ocean's surface through the layers of sand containing oil and other parts of the earth to the bottom of the well. *Downhole formations pressures* in the well are never known with certainty. If the drill bit enters what is called an "over-pressured" formation, that formation overpressure will exceed the weight of the fluid column in the well and begin to push fluid up the well. That phenomenon is called a *kick*.

What is in the well that is effectively vomited upward by this pressure? Mud. On a rig like *Deepwater Horizon*, you have people who specialize in making sure that the mud has the precise composition to keep the oil and gas down where it belongs. It's maintained by skilled people 24/7.

On *Deepwater Horizon*, there was no problem with the mud. In fact, it should be noted that two of the people on duty who were killed when the blowout occurred were mud engineers: Gordon Lewis, 28, and Keith Blair Manuel, 56. They were doing their jobs, as they did every day, to try to mitigate dangers related to oil and gas pushing upward.

Kicks are common with drilling oil and gas wells, so all drilling rigs have equipment installed and tested called blowout preventers (BOPs) to stop unexpected fluid flow. The kick is caused by an imbalance in pressure; the blowout preventers are then closed to seal the well while the balance is restored. It's a routine sequence of events. But what if the blowout preventers do not stop the unexpected fluid flow—the kick? That is when a disaster is likely to occur, and the disaster is called a blowout.

There is a huge difference between an oil and gas well kick versus an oil and gas well blowout, that is, the uncontrolled release of crude oil and/or natural gas from an oil well after pressure control systems have failed. A well kick is benign; it will do no damage if it is contained by the mechanisms that compose the BOP stack. But a benign kick quickly turns into a disaster when the BOPs fail.

On April 20, 2010, the drilling rig crew observed a well kick that they expected they could handle and close in without problems. Drilling rig crews are trained to recognize well kicks and trained to kill them, and then they move forward drilling the well. The April 20 Macondo well kick wasn't a routine event with a standard ending, however. It rapidly evolved into a catastrophe because the BOPs did not function correctly.

At 9:14 p.m., the rig crew circulated the mud out of the well to replace it with seawater, which is much lighter, because they were preparing for a temporary abandonment of the Macondo well. Rigs like *Deepwater Horizon*, which was designed to drill for oil and not extract it, abandon wells regularly. They determine what kind of production yield the company might get from a site and then move on. For that reason, the process of temporarily abandoning a well was familiar to the *Deepwater Horizon* crew. But this time, a few parts of the process deviated from what had occurred before. For one thing, the request to do a temporary abandonment of the well was made by BP corporate on April 16 and approved *minutes* later by an MMS official—a fact covered in the discussion of Mistake #2. But there was a more serious issue related to the mud. As part of this abandonment process, displacing the mud between the drill-pipe depth and the blowout preventers with seawater was something that BP engineers thought would make it more likely be successful in placing a temporary abandonment cement plug below the well head. (Some engineers believe that seawater is a better fluid environment to achieve a successful cement plug, while others do not agree.) Cementing is part of the proper procedure for abandoning a well. The rest of this abandonment of Macondo did not follow all recommended practices, however; I explain the significance of this in Chapter Two. The plan was standard—seal the well so BP could move on to another drill site—but the execution deviated from what anyone on the rig had done before.

For just a moment, rise above the *Deepwater Horizon* rig and look at offshore drilling in the Gulf of Mexico from the perspective of a jet passing over the area. At its widest point, the Gulf stretches a little more than 932 miles and encompasses more than 600,000 square miles, so it roughly the size of Iran. (Interestingly, Iran is where BP got its start when a group of British geologists discovered oil there.) In this vast area of water, according to the United States Energy Information Administration (EIA), there were twenty offshore drilling rigs in the Gulf of Mexico when *Deepwater Horizon* was afloat.

Now zoom down to the water and plunge beneath it to the seabed. If you could explore every square foot of the ocean floor, you would find more than 3,000 active wells in the Gulf according to the United States National Oceanic and Atmospheric Administration (NOAA). But you would find far more inactive wells than active ones. As of July 2010, just after the blowout, there were more than 27,000 unused wells.[2] Macondo was to be one of 600 abandoned, at least on a temporary basis, by BP alone.[3]

At 9:14 p.m., on that warm, calm evening, no one on the drilling rig except for two BP senior level managers and one Transocean senior level manager visiting the rig, were aware that in April 2006, BP and Transocean senior level managers in Houston had created a ticking time bomb below the *Deepwater Horizon* and on April 15, 2010, the BP Macondo drilling team made BP policy-deviation decisions that initiated the triggering device on the ticking time bomb. Long term, deliberate senior level management decisions at BP and Transocean, compounded with the implementation of the temporary abandonment decision, triggered the time bomb. As the story develops, I will get specific about what these managers and their bosses in the executive suites decided that contributed to the cascading failures that led to the blowout.

From 9:14 until 9:27 that evening of April 20, there were no displacement anomalies in the well; the displacement of mud (replacing it with seawater) was proceeding as planned. Just after 9:27, things started to change drastically onboard the *Deepwater Horizon* drilling rig floor.

After 9:27, the driller and toolpusher noticed pressure and flow changes on rig floor gauges. A toolpusher is a foreman who supervises drilling operations. The driller is a team leader in charge of the process of well drilling.

At 9:30, a high-pressure pop-off valve released on main rig pump #2, so the driller shut down all main mud pumps. In response, he and the toolpusher discussed what could cause the pressure changes.

The minutes ticked on and, even by 9:39, no one on the rig floor was seriously concerned that the well was kicking; in fact, the toolpusher left the rig floor to go down to the mud pump room to check pump #2. As soon as the toolpusher left, gas in the riser began to expand faster, pushing mud out of the riser—the 5,000-foot pipe connecting the rig to the BOP.

Thirteen minutes after the sudden increase in pressure—at 9:40—mud started to bubble out of the well and flow onto the rig floor. As soon as the driller saw mud spewing on the rig floor he activated the "marine riser diverter system." This system first closed in the mud flow from the riser and then diverted the mud to a mud-gas separator. The separator is a mechanism that captures and separates gas within the mud; it then forces it through a line and vents the gas safely. The toolpusher had hurried back to the rig floor and at 9:41, he was actively addressing the problem as well. The first thing he did was attempt to close the upper annular blowout preventer.

There are two types of BOP valves: annular and ram. The annular system is primary; there is an upper annular unit *and* a lower annular unit. An annular mechanism is a kind of rubber doughnut that squeezes tighter and tighter until it creates a seal. The toolpusher tried to avert further effects of the kick by activating the annular system.

Despite closing the annular system, the toolpusher was unsuccessful in getting the upper annual preventer to stop the flow of mud out of the well. As mud continued to spew out of the well, the toolpusher called the on-duty BP well site leader, the driller notified the maritime crew in the Central Control Room/Bridge, and the assistant driller called the senior toolpusher in his bedroom to inform him that the rig had taken a well kick. The maritime crew is a relatively small cadre of the professionals on the rig, but they are vital because the rig is a floating vessel. It would be up to the senior member of the maritime crew, Captain Curt Kuchta, to make the call that the rig should be disconnected from the riser and turned into a true sailing vessel. Of course, even with the authority to do it, he still needed the electric power to disconnect.

When the annular system failed to shut-in the well kick, at 9:46, a rig crew member closed the BOP stack's variable bore rams, VBRs. "Variable" means that the equipment can seal around different pipe sizes. As opposed to the rubber cap of the annular system, this is a steel-bodied clamp on the pipe. In the *Deepwater Horizon* BOP stack, there were two sets of clamps. After that is a blind shear ram, which is designed to both clamp the pipe and cut it below the clamp.

The upper VBR appeared to seal the well kick briefly, but soon mud continued to flow out of the well. At 9:47, twenty minutes after the kick and two minutes before the explosion, a crew member closed the next VBR; it also did not seal the well.

At 9:49, there was an explosion on the rig that killed all power on the rig floor. As soon as power was lost, the last-option "deadman" system on the subsea BOP should have automatically activated. The Automatic Mode Function (AMF)/deadman system comes into play only when all other systems have failed. In this case it could not shear the pipe and seal the well kick because one battery for the deadman was dead and the other—even if it had some life—could not function because of mis-wiring in the pod where it was encased. The batteries had not been

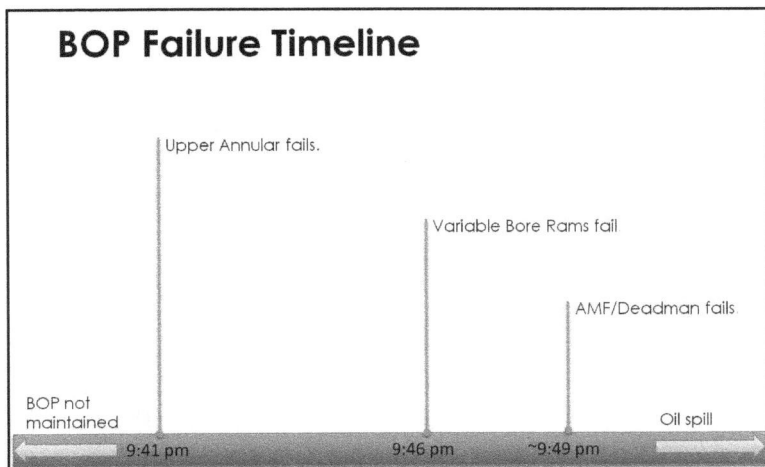

BOP Failure Timeline

Upper Annular fails.

Variable Bore Rams fail

AMF/Deadman fails

BOP not maintained

Oil spill

9:41 pm   9:46 pm   ~9:49 pm

changed in two-and-a-half years. The Macondo well kick turned into an out-of-control blowout.

The cascading effect with the BOP system failures occurred so rapidly that most people onboard were completely unaware there was anything to be concerned about until they heard a pervasive hissing noise followed by explosions, flames, and a total loss of power.

At 9:56, there was an attempt by a BOP engineer and the BP well site leader to activate the Emergency Disconnect System (EDS), which separates the drilling rig and marine drilling riser to allow the *Deepwater Horizon* to float away from the well kick. Without power, the EDS failed; the rig did not disconnect.

Here's where we are: On the *Deepwater Horizon*, there were two annular blowout preventers, two solid body variable bore rams (also blowout preventers), one blind/shear ram (another blowout preventer), an Emergency Disconnect System (to allow the rig to float away), and an AMF/deadman system (the final blowout preventer, designed to cut through the pipe so oil/gas would stop moving toward the rig). All of the BOP systems were activated and all failed. The choice for all those onboard was simple at that point: Leave the rig or die in a blaze.

The sequence of events related to the BOPs that culminated with the failure of the AMF/deadman—the "last hope"—may be clearer if you look at the drawing of a BOP stack.

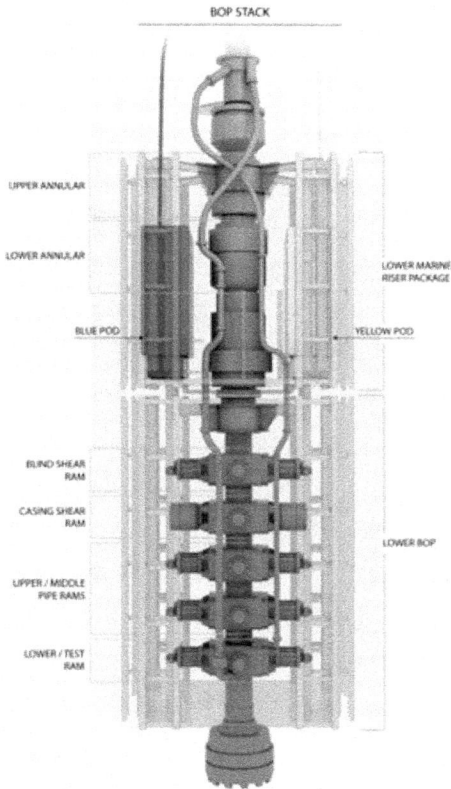

- The upper annular preventer failed.
- The upper and middle ram preventers failed.
- The blind shear ram failed.
- The deadman's ability to shear the pipe could not be triggered because it had no power.

To clarify this last point, the AMF/deadman relies on two redundant control pods. Under normal operations, the pods are powered through AC cables from the surface. In the event of a loss of power from the surface, the power supply for each of the pods is maintained through batteries in each pod. Each one functions independently, and each has its own power supply and batteries. Each pod includes solenoid valves, which are devices that open and close in response to electrical signals. The solenoids are designed to communicate with the BOPs and trigger the delivery of pressure sufficient to enable the AMF to sever the pipe.

It seems inconceivable that one system after another would fail, doesn't it? If you consider that the **BOPs were required by law to be inspected every three to five years**, according to both the manufacturer and the federal government, the disaster seems more predictable. **BP had never inspected the BOP stack since the date of its installation, December 13, 2000.** That period of nearly ten years, therefore, was at least double the length of time in which an inspection should have occurred.

And consider that something like a mis-wired solenoid should have been detected in routine inspection and tests at the time of installation. As we go through this story, you will see countless instances of basic failures like this. How many people signed off on pieces of equipment and approved processes that they never scrutinized? How many federal officials assumed that the corporate engineers had a grip on issues and could be trusted?

What the company did was a violation of the US Code of Federal Regulations (CFR), as well as the company's own maintenance manual. 30 CFR 250.446 directly references the American Petroleum Institute (API) Recommended Practices regarding BOP maintenance, thereby making those practices the "law of the land." They state unequivocally: "After every 3-5 years of service, the BOP stack, choke manifold, and diverter components should be disassembled and inspected in accordance with the manufacturer's guidelines."[4]

Your list of suspects now expands greatly. As I will cover in the upcoming chapters, there were people on the rig, people who had been on the rig, and people sitting in comfortable BP offices who had—or most likely had—knowledge that the BOP was way out of federally mandated maintenance and inspection compliance. Some of those people gave their consent to foregoing the required inspection that would have cost the company a minimum of $15 million in lost drilling time and hard costs of conducting the inspection. And even after there was concern expressed about the upper and lower annular compliance, just days before the disaster on April 15, email traffic indicated that executives thought there a low risk of having multiple failures.[5] To recap the way we introduced Mistake #1, they were hoping the BOPs would work instead of taking action to ensure their integrity. The official word was—and it certainly sounds cavalier in hindsight—"we don't want to the change the annulars. . ." The official corporate response was "BP accepts responsibility if both annulars were to fail."[6]

When "accepting responsibility" is a dollars-and-cents proposition, the big challenge may be how to show the loss on the books. In this case, "accepting responsibility" had graver consequences. Someone gave the okay that put people on the rig in harm's way. In fact, many people during the four or more years the BOPs were out of compliance had to sign off on total or partial inspections and maintenance of BOP equipment.

In the aftermath of the blowout, *Fortune* magazine published an investigative piece called "BP: 'An accident waiting to happen'". The article revealed "a saga of hubris, ambition, and a safety philosophy that focused too much on spilled coffee and not enough on drilling disasters."[7] But what *Fortune* concluded early on was not fully verified until more *than five years later* during my trial.

In the decade before the *Deepwater Horizon*, BP had a history of serious accidents. Each time its CEO vowed to avoid a future disaster. In 2000, after a string of fires and equipment failures, CEO John Browne announced plans to "renew our commitment to safety." In 2005, after a horrific explosion killed 15 people at BP's

Texas City refinery, he swore there'd be "no stone left unturned" to investigate what happened and correct any safety issues. In 2007, after being named Browne's successor in the aftermath of more problems, Tony Hayward promised to focus "like a laser" on safety—only to oversee the worst oil spill in history.[8]

The ironies in the statements of BP's most senior executives were not only painful, but also deadly.

Mistake #2: Fueling a culture of collusion between corporate and government employees

This mistake involves a significant accusation of wrongdoing on the part of both company executives and United States government employees. There is a foundation for the accusation, however, and it is well documented by insiders at the time and by credentialed outside analysts who examined the tragedy later. Upfront, it is important to know that the Minerals Management Service (MMS) of the U.S. Department of the Interior (DOI) was the federal agency primarily responsible for regulating the safety of offshore drilling at the time of the *Deepwater Horizon* tragedy.

Let's first look at a couple of results of the relationship between a key MMS employee and BP, and then see how this fits into the larger picture of a culture tainted by too much trust and too little oversight.

Frank Patton was the district drilling engineer with the MMS permits section. He would be the person to approve exemptions in testing procedures and modifications to the Macondo well. Exemptions to federal regulations, such as allowing testing of BOPs to 50 percent of rated working pressure as opposed to the specified 70 percent, would have to be approved by Patton—and he did approve such a request within minutes of receiving the email request according to email timestamps. BP's justification for wanting to test to a lower pressure was ostensibly to make the blowout prevention equipment last longer. Of course, for most engineers, this logic belies the reason for the test in the first place. It's like testing a swing that should hold a teenager by putting a toddler on it.

Patton had a history of rubberstamping requests related to *Deepwater Horizon*. On April 16, four days before the blowout, he got an email request to approve a temporary abandonment of the Macondo well. It was sent at 3:25 on a Friday afternoon with a request to approve it "today" because operations to shut down the well were scheduled to begin Sunday. Patton returned his approval of the request twenty-two minutes later, at 3:47.[9]

A *New Orleans Times-Picayune* headline from May 11, 2010 captures the most egregious action Patton took, however: "Updates from oil rig explosion hearings: MMS engineer admits he approved blowout preventer without assurances it would work."[10]

MMS was plagued by criticism and controversies. The problems precede the *Deepwater Horizon* blowout by years, but they came to a head with revelations related to it. That ultimately led Secretary of the Interior Ken Salazar to break MMS into three separate divisions—a decision he announced within one month after the *Deepwater* tragedy. Salazar aimed to stop what former Inspector General of the Interior, Earl Devaney, described as "a culture of ethical failure" nearly two years before in a September 9, 2008 report to then Secretary Dirk Kempthorne. Salazar did it by making sure that a single organization was no longer responsible for leasing and royalty management—revenue-focused activities—and issues of safety and environmental protection. The conflict of interest should have been obvious long before the April 2010 BP blowout.

After two years of investigations, Earl Devaney concluded that some MMS staff "accepted gifts with prodigious frequency"[11] He even said nearly *one-third* of the entire Royalty in Kind (RIK) staff "socialized with, and received a wide array of gifts and gratuities from, oil and gas companies with whom RIK was conducting official business." We are just scratching the surface, however. Devaney went on to explore the debauchery:

> We also discovered a culture of substance abuse and promiscuity in the program—both within the program, including a supervisor, _____, who engaged in illegal drug use and had sexual relations

with subordinates, and in consort with industry. Internally, several staff admitted to illegal drug use as well as illicit sexual encounters. Alcohol abuse appears to have been a problem when RIK staff socialized with industry. For example, two RIK staff accepted lodging from industry after industry events because they were too intoxicated to drive home or to their hotel. The same RIK marketers also engaged in brief sexual relationships with industry contacts. Sexual relationships with prohibited sources cannot, by definition, be arms-length.[13]

This is not meant to suggest that Frank Patton was among the MMS offenders referenced in the report; there is no indication he was. But as the agency's New Orleans District drilling engineer, it was his job to exercise oversight rather than let industry requests slide through the system. In an interesting outcome, Patton still works for the federal government, becoming a petroleum engineer with the Bureau of Safety and Environmental Enforcement (BSEE), the successor organizations to MMS responsible for environmental protection and human safety.

Mistake #3: Cherry picking which policies would be adhered to

A fat book known as the *Drilling and Well Operations Practice* (DWOP) is the manual issued to selected BP employees. It's labeled "highly confidential" on each page and is so closely guarded that authorization for issue to an employee must be given in writing—with a signature—from senior BP officials such as the Technology Vice President and a Vice President of Drilling and Completions. The DWOP guides the user in enforcing standard procedures and evaluating significant risks related to Engineering Technical Practices (ETPs) such as those that applied to *Deepwater Horizon*. Paragraph 1.4 at the very beginning of the manual states, "All staff and contractor personnel in managing BP drilling and well operations shall be knowledgeable of all elements of this practice and associated ETPs and are responsible for conformance."[14]

The practices detailed in the book number in the hundreds. For example, in one particularly relevant section, the DWOP states: "Annular BOPs shall be tested to a maximum of 70% of rated working pressure." This is one of those policies that wouldn't necessarily be glossed over, but it might be the subject of an email seeking an exemption, as I just discussed in Mistake #2 when an exemption was granted to test at 50 percent rather than 70 percent.

The harsh reality is that the privileged few with access to the DWOP—and I was one of them—could conceivably cherry-pick which practices they felt should be honored to the letter, which could be honored in spirit, and which could be ignored until it was convenient.

Most people have a hard time with the concept of absolutes because absolutes rarely match well with the realities of daily life. In the context of the *Deepwater Horizon* tragedy, several companies, particularly BP, had faced the economic realities of conforming to absolutes as they pertained to maintenance, inspections, and other safety conformance issues. All I am asserting here is that they strayed too far from compliance and that turned out to be a life-threatening mistake. So regardless of the stipulation cited above that "All staff and contractor personnel . . .shall be knowledgeable of all elements. . .and are responsible for conformance," that condition was not executed in an "absolutist" manner—or anywhere close to it. The unfortunate outcome was the blowout that endangered 126 human lives, killed eleven people, and did incalculable damage to the environment. The enormity of the event shocked the world; the fire that resulted could be seen by the Coast Guard rescuers about fifty miles away as they flew their helicopters toward the burning rig. A blowout on an offshore rig like *Deepwater Horizon* is a situation in which absolutes dominate.

In the next chapter, I'll tell what I did on *Deepwater Horizon* and how that figured into the jobs and objectives of both the people around me and the executives I reported to.

CHAPTER TWO:

4 ½ DAYS

I was one of 126 people on *Deepwater Horizon* on the night of the blowout. My presence there was a combination of serendipity and trying to be a nice guy.

We were having a modification done to our production facility at *Thunder Horse*. Like *Deepwater Horizon*, *Thunder Horse* PDQ is a semi-submersible oil platform, but unlike it, *Thunder Horse* does more than explore. The PDQ identifies it was a Production and oil Drilling facility with crew Quarters. *Thunder Horse* is a joint-venture of BP and Exxon Mobil and the largest offshore installation of its kind in the world.

We had a planned cessation of drilling operations. The company needed the bed space for all of the maintenance and construction people who were handling the modification, so we had to move elsewhere. Temporarily, I was transferred off the rig and into the Houston office. A little while after I got there, someone from the company came in and asked me if I'd relieve Ronnie Sepulvada on *Deepwater Horizon*. He had to leave the rig to go to well control school for five days. I said, "Sure!"

That cooperative and easy decision ended up becoming the epitome of "no good deed goes unpunished."

I had been there just four and a half days when the rig was consumed by flame and smoke.

I was there to finish the temporary abandonment of the Macondo well. I had no idea that the BOPs were out of maintenance and inspection compliance until nearly a year later when I read the Chief Counsel's report related to the initial hearing about the blowout. But it stands to reason that someone on the rig, as well as many people off the rig, knew the BOP equipment was out of compliance.

When a disaster of this magnitude occurs, people knowledgeable about offshore drilling as well as people who know nothing at all about it are likely to have the same general speculations about what caused it. It was either equipment failure or negligence on the part of rig workers or their employers—or both. We now know that equipment failure resulting from negligence caused the *Deepwater Horizon* blowout, but exactly whose negligence?

To take a few steps toward the answer, let me introduce you to people on board the rig and explain what jobs we were doing in the days that led up to the blowout.

The *Deepwater* Crew

In addition to the rig crew, other people with central roles in the story of the disaster are the BP and Transocean Macondo operations and engineering teams, the rig auditors, and federal government employees charged with enforcement of regulations and issuing permits and waivers. In those final days of *Deepwater Horizon*, those of us on the rig were just doing our usual jobs until the evening of April 20, when the impact of what all the other key players had done—or not done—changed our lives forever. The identification and involvement of those individuals is covered in Chapter Five.

Of the 126 people on board, seventy-nine were Transocean employees—the drilling crew—seven of us worked for BP, and the other forty people were contractors from companies like Halliburton and Schlumberger's M-I SWACO division, which supplied materials and specialized services to BP and *Deepwater Horizon*. As you read through the brief descriptions of what many people on an offshore rig do, keep in mind that we worked in two shifts, so you might think of roughly sixty people doing these jobs at any given time, with the other sixty sleeping, eating, or having a conversation with a loved one onshore. The two-shift schedule is common on an offshore drilling rig due in large part to continuous 24-hour work and space constraints.

Jimmy Harrell was the most senior person on the rig as long as it was attached to the well head, as opposed to being a floating vessel. Jimmy was the offshore installation manager (OIM) for Transocean, that is, the top drilling official. The OIM can override anyone's decision on the rig or "in town," meaning the onshore location where orders come from. The first action might be to call

Transocean and protest: "I'm not doing that." Or if I had said something to him that he didn't like—and let's say I was speaking as a senior BP person on the rig—he could say "no." And this does happen. The OIM is the top dog.

For those who have seen the movie, that means that when Kurt Russell, who portrayed Jimmy Harrell, took exception to decisions made about cement in the well casing and other substantial issues, he could have called the shots. Kurt Russell plays a hero who was overridden and ultimately victimized by BP employees' decisions. The actual chain of command put Jimmy Harrell in charge of all the decisions made on *Deepwater Horizon*.

The OIM has maritime training, but whenever the rig is a maritime vessel, that is, not attached to anything and floating freely, then the captain is in charge. With a disconnection from the well, which is what we were about to do with the abandonment of Macondo, Captain Curt Kuchta, would have taken over as the top decision-maker onboard unless his replacement had arrived. Captains work two weeks on and two weeks off and are supported by a small maritime crew that includes people such as a ballast control operator, who is responsible for controlling the rig's trim, draft, and stability.

The rest of us had jobs like driller, assistant driller, mudman, toolpusher, roughneck, crane operator, roustabout, electrician, rig mechanic, and engineer. Roughly two-thirds of the population of the rig belonged to the Transocean drilling crew.

A roustabout would be a good starting job. Theoretically, it could be someone who simply knew a lot about tools and how to use them. When I worked in Alaska on the North Slope drilling rigs, we had roustabouts who were just two years out of high school and making $80,000 a year.

> Roustabout: Any unskilled manual laborer on the rigsite. A roustabout may be part of the drilling contractor's employee workforce, or may be on location temporarily for special operations. Roustabouts are commonly hired to ensure that the skilled personnel that run an expensive drilling rig are not distracted by peripheral tasks, ranging from cleaning up location to cleaning threads to digging trenches to scraping and painting rig components. Although roustabouts typically work long, hard days, this type of work can lead to more steady employment on a rig crew.[1]

Roughneck, also known as a floor man, is a job that a roustabout might work up to. There might be three or four on the rig floor at one time, doing things like connecting pipes.

> Roughneck: A floor hand, or member of the drilling crew who works under the direction of the driller to make or break connections as drillpipe is

tripped in or out of the hole. On most drilling rigs, roughnecks are also responsible for maintaining and repairing much of the equipment found on the drill floor and derrick. The roughneck typically ranks above a roustabout and beneath a derrickman, and reports to the driller.[2]

The youngest man to die in the blowout was the lead roughneck, twenty-two-year-old Shane Roshto from Liberty, Mississippi. During the four years he had served onboard *Deepwater Horizon*, he had worked his way up to that spot. In tributes to him after his death, survivors noted that he had a strand of steel cable, like the kind used on rigs, embedded in his wedding band.[3] Other floor men who were casualties of the disaster were Karl Kleppinger, Jr., a veteran of Operation Desert Storm, and Adam Weise, a Yorktown, Texas native who had used his earnings to buy a black, four-wheel-drive diesel pickup he named "The Big Nasty."[4] Weise was just twenty-four years old when the tragedy occurred.

A mudman ensures the properties of the Synthetic Oil Based Mud (SOBM) meet planned drilling requirements, monitors the composition every few hours and weighs it, and assists, at times, in cleaning the mud pits. The people in charge of the composition of the mud are known as drilling fluids specialists—*aka* mud engineers—and on *Deepwater Horizon*, they were employees of Schlumberger's M-I SWACO division. As I mentioned in Chapter One, both the senior drilling fluids specialist, fifty-six year old Keith Manuel, and the more junior specialist, twenty-eight year old Gordon Jones, were killed in the explosion. A twenty-seven year old mudman named Roy Wyatt Kemp also died that tragic night.

> Mud engineer: A person responsible for testing the mud at a rig and for prescribing mud treatments to maintain mud weight, properties and chemistry within recommended limits. The mud engineer works closely with the rig supervisor to disseminate information about mud properties and expected treatments and any changes that might be needed. The mud engineer also works closely with the rig's derrickman, who is charged with making scheduled additions to the mud during his work period.[5]

Mud provides hydrostatic pressure to prevent the liquids and gases in the well from pushing upward. You know you have a big problem if the mud pushes upward onto the rig floor—which is what people saw the night of the blowout. Another reason that SOBM is used in some environments is protection for workers. The toxicity of the fluid fumes is much less than that of an oil-based mud (OBM) in which the base fluid is a petroleum product. In an enclosed space, this becomes important.

Toolpushers keep the rig running and are also known as drilling foremen. A toolpusher is the person in charge of the drilling department who reports to the OIM. Back in my youth, I became the second youngest toolpusher on Alaska's North Slope.

> Toolpusher: The location supervisor for the drilling contractor. The toolpusher is usually a senior, experienced individual who has worked his way up through the ranks of the drilling crew positions. His job is largely administrative, including ensuring that the rig has sufficient materials, spare parts and skilled personnel to continue efficient operations. The toolpusher also serves as a trusted advisor to many personnel on the rigsite, including the operator's representative, the company man.[6]

Transocean had two senior toolpushers onboard *Deepwater Horizon* the night of April 20. One was Randy Ezell, and the other was Jason Anderson, who died after the explosion. Jason had been on that rig since it launched from a South Korean shipyard in 2001.[7] Randy had been on *Deepwater Horizon* for eight years.

Jason was not supposed to be on the rig that day. He had stayed a little longer to say goodbye to his long-time rig buddies after his replacement arrived.

Observations Jason had made to me earlier that day about a pressure reading we got during one of the tests are an interesting part of the discussion that follows in Chapter Three. Jason's death meant that I was the only one who was legitimately able to share the substance of our conversation, which had the potential to be critical in the investigation of the disaster. That didn't stop other people with bits and pieces of information from relaying what they thought they knew, however.

There were other jobs and other people, of course, such as crane operators like Aaron Dale Burkeen of Philadelphia, Mississippi who died that night at the age of thirty-seven. The explosion caused him to fall onto the deck.

Both a chief driller, Dewey Revette, and two assistant drillers, Stephen Ray Curtis and Donald "Duck" Clark, were also killed by the blowout. The commonality is that they were on the rig floor, most likely throughout the two negative tests covered in the next chapter. They would have been the ones who recognized the kick at 9:41 that night, acted to control it, and were in the heat of the disaster at 9:49.

Senior Transocean Crew

Jimmy Harrell
Offshore Installation Manager
6 years on *Deepwater Horizon* (30 years total)

Randy Ezell
Senior Toolpusher (Day)
8 years on *Deepwater Horizon* (33 years total)

Jason Anderson
Senior Toolpusher (Night)
9 years on *Deepwater Horizon* (14 years total)

Dewey Revette
Driller
7 years on *Deepwater Horizon* (29 years total)

Don Clark
Assistant Driller
9 years on *Deepwater Horizon* (14 years total)

Steve Curtis
Assistant Driller
8 years on *Deepwater Horizon* (8 years total)

Don Vidrine and I were also part of the *Deepwater* crew, but unlike the drilling team employed by Transocean and several of the contractors who reported to Schlumberger or Halliburton, we were BP employees. Don belonged to the BP Macondo operations and engineering team to which I had been temporarily assigned.

As well site leaders for deepwater drilling structures, Don and I coordinated drilling operations on rigs in the Gulf of Mexico. There are different types of rigs, but in the case of *Deepwater Horizon*, we were supervising on a mobile offshore drilling unit (MODU). Part of our job, the part we were doing on April 20, 2010, was conducting tests in support of cessation of drilling operations and temporary well abandonment.

Lead-up to the 2 Tests

When I arrived by helicopter on the rig around 8 a.m. the morning of April 16, the other rig supervisor, Don Vidrine, told me he would stick with the night shift since he had gotten used to it. I got acclimated and prepared to do tests that Ronnie Sepulveda had planned to do, consistent with common procedures on the rig. Our goal was to ensure there would be no leaks when we abandoned Macondo, which had proven to be somewhat problematic. *Deepwater Horizon* would disconnect, become a maritime vessel once again, and head to a new location to drill another well.

To understand the nature and significance of the tests done on April 20, you need to have a paragraph of background on where leaks could occur.

Macondo had been drilled to a depth of 18,360 feet and *production casing* had been run to a depth of 18,304 feet into the well. After drilling through rock formations to reach the reservoir below, the raw sides of the well need reinforcement. Production casing is tubing that is set inside the well to provide that protection. A dual float valve—a one-way valve, so fluid can go down the casing but not up the casing—had been installed in the casing at 18,115 feet. Cement had been pumped past the float valve.

BP engineers estimated that the four lowest casing joints, at depths of 18,115 to 18,304 feet, were full of cement, but they were not certain of that.

The tests that Don Vidrine and I were supposed to conduct were designed to make sure that the casing down to the float valve at 18,115 feet and all the way up to the BOP stack more than 13,000 feet above it were not leaking before the rig abandoned the well.

The day we did the tests looked like this for me:

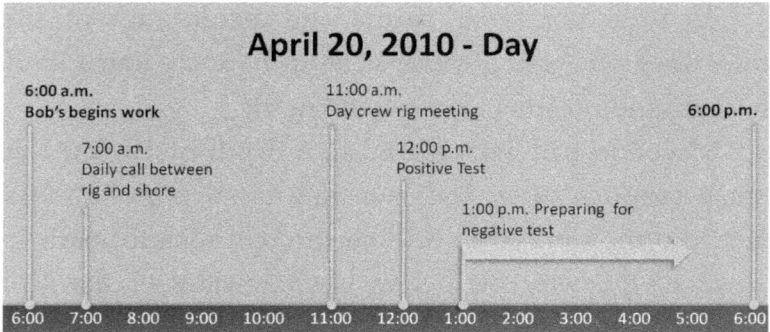

April 20, 2010 - Day

6:00 a.m.
Bob's begins work

11:00 a.m.
Day crew rig meeting

6:00 p.m.

7:00 a.m.
Daily call between
rig and shore

12:00 p.m.
Positive Test

1:00 p.m. Preparing for
negative test

6:00 7:00 8:00 9:00 10:00 11:00 12:00 1:00 2:00 3:00 4:00 5:00 6:00

In the positive test, fluid was pumped below the closed blind/shear rams (BSR) in the BOP stack. There was a cement plug on top of the dual float valve at 18,115 feet so pressure between that point and the BOP increased in the production casing while fluid was pumped into the well. When the pressure approached approximately 2,600 psi, pumping was stopped. The positive pressure build-up in the production casing was held for thirty minutes.

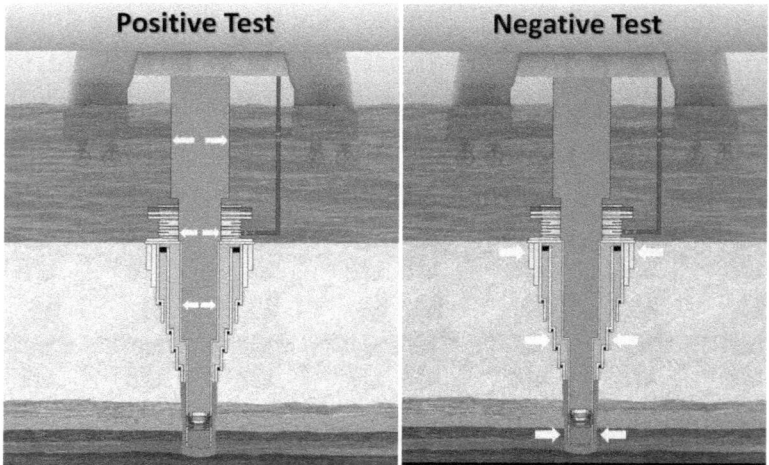

Positive Test | Negative Test

Because the pressure did not decline at all, we concluded the casing was not leaking.

After we completed a successful positive test on April 20, we did two negative tests—the source of intense scrutiny by all the sources of investigative reports related to the blowout and the subject of the next chapter.

First, let's skip ahead to my last few hours on board *Deepwater Horizon*.

The Final Hours

The question that many have explored, but no one can answer with certainty, concerns exactly what happened in the hours preceding the well kick at 9:41 that culminated in the blowout. Attempts were made by BP to create a direct cause-and-effect relationship between the negative tests and catastrophe, however, that makes no sense— and I explore that in Chapter Three. In essence, the tests were designed to evaluate whether or not the casing and cement at the base of it could withstand a certain pressure differential without leaking. We got the answer to that: no leaks. Issues that came up after that relate to actions after the tests.

For obvious reasons, the post-mortem could only do so much because of the destruction of the rig, oil pouring out of the well, and then "burial" of the well with cement. The ability to draw conclusions from science to solve any mystery associated with the *Deepwater Horizon* tragedy was severely limited by circumstance. One could argue (reasonably) that BP and Transocean executives were also not going to make an all-out effort to uncover all the facts after they went to such great lengths to construct their case against me and Don Vidrine.

Another critical chunk of time during my four-and-a-half-days on the rig was the eight-minute period between the moment the kick was detected and the blowout. I was asleep, but I will try to reconstruct what happened.

Let's start with the time and people associated with kick detection. Out of respect for the people who were doing their jobs, and had reasons to do their jobs a particular way, I'm going to refer to them by job title only. These men caused nothing, did nothing wrong, and died giving their best efforts to solve a horrible problem.

After reading all the investigative report sections concerning kick detection, I can't be a harsh critic of the driller and toolpusher because I know from *experience* that a kick is not always easy to detect. That said, some signs do point to the probability that they had a good opportunity to predict the kick at 9:30 p.m.—nineteen minutes before the rig explosion, or eleven minutes sooner than they detected it.

At 9:30, the driller shut-down the rig pumps to evaluate unusual pressure. Over the next nine minutes, he and the toolpusher discussed the situation and acted by bleeding drill pipe pressure. They did not conduct a flow check, though, and did not shut-in the well. This becomes significant if those actions would have contained the situation before reservoir fluids migrated above the BOPs and into the riser. Notice I say "if."

Hindsight is not always 20/20 because we'll never know how kick detection decisions were made on *Deepwater Horizon*. Unfortunately, the best witnesses aren't here to tell us. After the rig accident, analysts, so-called experts, and critics have looked at pressure charts, read or heard testimony and tried to evaluate why the rig crew didn't close the BOPs at 9:30 or even conduct a flow check at the time. Most hindsight observers have seen that the experienced driller detected a pressure anomaly at 9:30, took the action to shut-down the rig pumps, discussed his concern with the experienced toolpusher, and had an opportunity to at least flow check the well, but neither one did a flow check. In hindsight, the driller and toolpusher had a second opportunity to flow check the well due to a second pressure anomaly at 9:39, ten minutes before the rig explosion, but they did not.

Those same experts did some flow modeling, asserting that the Macondo well could have been flow checked and shut-in *before* reservoir fluids ever migrated above the BOPs and into the riser. It's a theory put forth by BP whose experts did the modeling, but keep in mind that the people with eyes and hands on the problem worked for Transocean, not BP.

There are many factors that probably affected their decision to not flow check the well or shut-in the well before mud overflowed onto the rig floor. A key one is this: The driller had observed and participated in two successful negative tests prior to the final displacement and the on-tour toolpusher had taken the operation lead and observed one successful and valid negative test, which confirmed well integrity prior to the final displacement. I am certain the driller and toolpusher had heard information that the casing cement job was good, and both had seen and participated in the successful negative tests, so each supervisor knew that the well was absolutely static and secure before the final displacement. They probably concluded they had no reason to go on "red alert."

In early 2011, the National Commission on the BP *Deepwater Horizon* Oil Spill and Offshore Drilling issued The Chief Counsel's Report, a 371-page document that examined technical, management, and regulatory findings related to the disaster. Appointed by President Barack Obama, the Commission Chairs were William Reilly, Administrator of the Environmental Protection Agency under President George H. W. Bush, and Senator Bob Graham (D-FL). Pages 165-190 of Chief Counsel Fred Hartlit, Jr.'s Report give the Commission's best efforts to provide a detailed look at what happened during the last hour and forty-seven minutes before the blowout. Based on the information in that Report, which is freely available online (www.wellintegrity.net/documents/ccr_macondo_disaster.pdf), I designed the following table. The descriptions will come alive after you become more familiar with the negative tests, but here they serve to set the stage for delving into the tests that BP pointed to as "failed"—thus justifying the Department of Justice's prosecution of me.

Time: hrs	Operation	Minutes from Explosion	Shut-in BOPs	Comments
8:02	Began second displacement	1 hr, 47 minutes	No	Taking on seawater into the pits "on-the-fly" (that is, directly from the ocean pump to the main rig pumps), could not accurately monitor flow into the well
8:34	While displacing well and riser	1hr, 15 minutes	No	Rig crew redirected flow into reserve pit, dumped sand trap into active pits, filled the trip tank – crew moving mud in the active pits while displacing
8:49	While displacing well and riser	1 hr	No	Rig crew redirected flow into another reserve pit. Actions consistent with cleaning pits.

8:50	While displacing well and riser	59 minutes	No	Via Transocean calculation: the well became underbalanced. BP calculated at 8:52
8:52	While displacing well and riser	57 minutes	No	Don Vidrine called Mark Hafle, BP's Senior Drilling Engineer. Hafle had Sperry Sun real-time data on his computer screen
8:59	Slowed mud pumps	50 minutes	No	Slowed mud pumps and rig crew dumped trip tank
9:01	Slowed mud pumps	48 minutes	No	When pump rate leveled out, drill pipe pressure increased slightly; the first indication something could be wrong.
9:08	Displacing at a slower rate	41 minutes	No	Drill pipe pressure had increased 100 psi but the increase would have been very subtle. Mudlogger did not detect the change.

9:08	Stopped mud pumps, to conduct sheen test	41 minutes	No	Flow chart indicates 1 minute of flow after pumps were stopped. Mud logger declared a "no flow" at the sheen test.
9:10	Diverted flow overboard	39 minutes	No	Mudlogger would not be able to monitor flow-out of the well. Also transferred mud between active pits.
9:08 – 9:14	During sheen test	35 minutes	No	Drill pipe pressure increased 250 psi. Driller, Assistant driller, mudlogger all missed the pressure increase.
9:17 – 9:20	Stopped main pumps to repair pump #2	29 minutes	No	Blew a pop-off valve on pump #2, shut down pumps #3 and #4 to identify which pop-off blew. Left pump #1 pumping on boost line.

9:20	Resumed displacement	29 minutes	No	Randy Ezell called Jason on rig floor. Jason stated, "It's going fine. It won't be much longer, I've got this."
9:14 – 9:27	Displacing well and riser	22 minutes	No	"The data did not clearly reflect any anomalies," according to Commission experts (p. 180).
9:27 – 9:30	Displacing well, opened kill line	19 minutes	No	Drill crew opened the kill line, noticed 800 psi kill line pressure while drill pipe had 2,500 psi
9:30 – 9:35	Shut down rig mud pumps	14 minutes	No	Driller and toolpusher discussed pressure differential. Drill pipe pressure increased 550 psi with mud pumps shut down. By analysis hydrocarbons had not entered the riser yet.

9:36	Bled pressure off of drill pipe	13 minutes	No	Floorman bled down drill pipe pressure from 1,600 psi to 550 psi
9:37 – 9:39	Drill pipe pressure increased	10 minutes	No	Drill pipe pressure increased from 550 psi to 1,450 psi. Did not conduct a flow check. Toolpusher left the rig floor to go to the pump room. No kick concern on the rig floor.
9:40 – 9:41	Mud overflowed onto rig floor, crew diverted flow to the trip tank	8 minutes	No	Rig crew diverted flow to trip tank to monitor flow gain.
9:41	Shut-in annular preventer	8 minutes	Yes	Toolpusher back on rig floor, shut-in annular preventer and diverted flow to the mud gas separator (MGS)

9:41 – 9:47	Rig floor actions	2 minutes	Yes	Toolpusher called well site leader, assistant driller called senior toolpusher, driller called the bridge. The annular preventer did not seal the well; maximum drill pipe pressure just 1,200 psi.
9:47	Shut-in well	2 minutes	Yes	Someone either closed a VBR or increased annular closing pressure. Drill pipe pressure increased significantly. Neither action sealed the well.
9:49	First explosion	0 minutes		Well site leader and senior toolpusher coming to rig floor

CHAPTER THREE:

2 TESTS

When Interstate 35W in downtown Minneapolis collapsed in 2007 and thirteen people were killed, should we have blamed the cars and school bus on the bridge at the time? Any reasonable person would say that doesn't make sense. The bridge over the Mississippi River was built to handle traffic; drivers aren't at fault for poor road maintenance. This analogy to the Macondo disaster isn't far-fetched. The negative flow tests initially blamed for the blowout are stresses that the systems onboard *Deepwater Horizon* should have been able to handle and, if not handle, then respond to without breaking down.

When systems or structures built to take significant stress fail completely, it is logical to blame those responsible for maintenance. But that simple logic escaped the Department of Justice when BP pointed to our negative tests as the cause of the tragedy. DOJ bought the bogus story, spotlighted to take attention off deferred maintenance of the BOP equipment—and it wasn't until my trial six years later that they had to publicly admit their mistake.

But there was a good reason why BP was able to get away with the pretense and it had to do with the BP Macondo drilling team's design for the tests rather than the execution.

The Why and What of Negative Testing

In a negative test, we lower the pressure to make sure that the production casing and at times, the cement at the base of it, can withstand a certain pressure differential without leaking. The objective is to simulate actual events that would occur during the well abandonment.

To do a negative test, we pump seawater through the drill pipe below the BOP stack and close the BOPs. Because seawater weighs far less than the mud—8.6 pounds per gallon versus 14 pounds per gallon—the displacement reduces the hydrostatic head. Mud provides hydrostatic pressure to prevent the liquids and gases in the well from pushing upward, so displacing the mud with seawater in a section of the casing allows the system to go into an underbalanced state. That condition enables us to check for leaks.

The Macondo drilling team changed negative testing procedures three times in the last eight days of the Macondo well. The first change improved the safety of the negative test, but the final two plans were changed without careful engineering analysis and risk assessment.

The Safe Test Design

As of April 12, 2010, there was no plan to conduct a negative test included in the approved well plan. Rig supervisors advised the Macondo well managers and engineers that the planning team neglected to include a negative test. By April 14, the Macondo engineering team had revised the well plan and created a carefully engineered and safe negative test plan; they sent it to the rig. In brief, here is what they proposed:

- Run the drill pipe into the well to a depth of 8,367 feet.
- Pump a 300-foot, open-hole temporary abandonment (TA) cement plug from 8,367 feet to 8,067 feet in the mud (SOBM) in the hole. Pull the drill pipe up to 6,000 feet. Allow the cement time to harden.
- Conduct a negative test at the BOP (BOP depth ± 5,000 feet) by displacing the kill line from 14.0 pounds per gallon (mud) with 6.8 pounds per gallon base oil. This type of negative test would create 1,872 psi negative pressure in the well. This is a safe level.
- Displace the mud in the well and riser from 6,000 feet with seawater. This would create 1,685 psi negative pressure.
- Pull out of the hole, pick-up the *lockdown sleeve*, run in the well with the lockdown sleeve with the drill pipe below the sleeve for extra weight,

and set the lockdown sleeve in the wellhead. (A lockdown sleeve is equipment that is installed in the wellhead to guard against uplift forces.)

As designed, this was the negative test sent to the rig on April 14, 2010, six days before the blowout. To summarize: The negative test plan was designed to comply with BP drilling standards and was a safe plan for two key reasons.

The negative test conducted at the BOP using base oil created a higher negative pressure in the production casing than would be created while displacing the mud out of the casing and marine riser.

More importantly, this plan directed the rig supervisors and rig crews to pump the temporary abandonment cement plug into the production casing and allow enough time for the cement to harden before displacing the casing and marine riser from heavier weight mud to lighter weight seawater. The temporary abandonment cement plug would act as a *second barrier* in the well while displacing to lighter fluid in the well.

The lockdown sleeve would then be set. In terms of function, the lockdown sleeve is like the wire cage over a champagne bottle. The wire cage prevents the champagne cork from popping out of the bottle if the bottle is shaken and the gas inside is excited.[1]

The negative test plan of April 14 met the criteria that a test must simulate actual events—and it was proven to be safe.

Unfortunately, the plan changed.

The following chart from the Chief Counsel's Report referenced first in Chapter Two in relation to kick detection visually presents the differences in versions of the tests as the consideration of cost and time savings overtook safety concerns.[2]

April 14 Morel Email	April 15 Well Plan/ April 16 MMS Permit	April 20 Ops Note	April 20 Actual Procedure
Run in hole to 8,367'	Negative test to seawater gradient (with base oil to wellhead)	Trip in hole to 8,367'	Trip in hole to 8,367'
Set 300' cement plug in mud **BARRIER**	Run in hole to 8,367'	Displace mud with seawater from 8,367' to above wellhead (BOP)	Displace mud with seawater from 8,367' to above wellhead (BOP)

Negative test with base oil to wellhead	Displace mud in well and riser from 8,367' with seawater	Negative test with seawater to depth 8,367' rather than with base oil to wellhead	Negative test with seawater to depth 8,367' rather than with base oil to wellhead
Displace mud in well and riser from 6,000' with seawater	Monitor well for 30 minutes/ conduct second negative test	Displace mud in riser with seawater	Displace mud in riser with seawater **BLOWOUT**
Set lockdown sleeve	Set 300' cement plug in seawater **BARRIER**	Set 300' cement plug in seawater **BARRIER**	
	Set lockdown sleeve	Set lockdown sleeve	

The Dangerous Test Design

The plan approved by MMS engineer Frank Patton the same day I came on board *Deepwater Horizon* had some inherent flaws. As I mentioned in Chapter One, Patton's approval of the new plan was given twenty-two minutes after he received the email requesting an okay. In my opinion, that's not enough time to make all the appropriate calculations. There's no evidence that he even did calculations, so this could be an automatic approval of a faulty design.

This second-to-last negative test plan does not meet the requirement that a negative test must meet or exceed the pressures created while executing the actual event. The April 15-16 negative test called for a negative test at the BOP to a seawater gradient which would create 1,404 psi negative pressure. After the negative test at the BOP, the drill pipe would be run to a depth of 8,367 feet and the well displaced to seawater which would create 2,350 psi actual negative pressure. A critically important fact is that BP management with oversight in engineering decided to set the temporary abandonment cement plug—the second barrier in the casing—*after* displacement of the lighter weight fluid in the casing and marine riser.

The final negative test plan called for use of the BOP equipment in a manner for which it was not designed. As noted above, it also directed the rig supervisors and crews to set the temporary abandonment cement plug, the second barrier in the casing, after the casing and marine riser was displaced with lighter weight fluid. The plan is outlined in an "Ops Note" in the above chart.

- Run drill pipe in the well to a depth of 8,367 feet.
- Displace mud with seawater down the drill pipe, up the drill pipe in the casing annulus, stop pumping once the seawater is above the

BOPs, and close the annular preventer rubber.
- Negative test by using the kill line. Observe for flow for 30 minutes.
- Displace the mud with seawater in the marine drilling riser.
- Pump the 300-foot cement plug from 8,367 feet to 8,067 feet.
- Set the lockdown sleeve using the drill pipe as the weight below the sleeve.

The Macondo well kicked during the final displacement of the marine drilling riser with seawater. The BOP equipment failed to seal the kick and therefore the well kick developed into a blowout.

The challenge we had with the final negative test ordered by BP for the Macondo site was that the plan deviated sharply from anything BP and Transocean supervisors had done before. In fact, the directive from BP to create 2,350 psi negative pressure in the production casing had never been done before.

The Macondo drilling team quickly went from a safe negative test plan sent to the rig on April 14, to a dangerous negative test due to one change. The Macondo team decided, without careful engineering and risk assessment, to set the temporary abandonment cement plug in seawater rather than mud, per the safe, first negative test plan.

The mud was much heavier than seawater, so the Macondo well was negatively pressure tested to ensure well control prior to displacing the casing and riser volume to lighter weight seawater. That describes part of the design of the safe April 14 negative test.

To implement the negative test procedure approved by MMS on April 16, the well would become hugely underbalanced to formation pressures when the mud was displaced out of the well at a depth of 8,367 feet. Additionally, this procedure did not meet the requirement of a valid negative test. The Macondo team called for a negative test at the BOP that would create 1,404 psi negative pressure, next the procedure called for displacing the well and marine drilling riser to seawater which would create 2,350 negative pressure. Negative tests are supposed to simulate what will occur during the actual event. The April 16 version of the negative test at the BOPs would not do that. Suddenly, with the blessing of the MMS, we had a test design of questionable validity that pushed the limits of equipment integrity—with BP standing by the decision to proceed.

Why?

The test was designed in this manner theoretically to save an estimated six to eight hours. A manager in Houston believed it is more effective to put a cement plug into seawater than it is to put it into mud. That may or may not be true; however, some expert cementers would argue that setting an open cased-hole cement plug in SOBM delivers better results. The test procedure was therefore changed without careful engineering analysis and risk assessment.

At that end-stage of the Macondo exercise, every hour was critical to the company because *Deepwater Horizon* was already forty-three days behind schedule and $40 million or more over budget. Just the cost of leasing the rig from Transocean was $500,000 a day. The rush to save time proved unwise, though.

The Infamous Tests: Part I

The first negative test was conducted from 5:18 to 5:35 p.m. on April 20. When I arrived on the drilling rig floor at 5:18 p.m. and entered the driller's cabin, I said "Hello" to everyone and asked enthusiastically: "Are you all ready to conduct the negative test?" The senior toolpusher on duty said, "Yes, we are!" I looked out toward the rig floor and observed that the rig crew had arranged the piping to conduct the negative test via the drill pipe.

At the morning rig crew meeting, held at 11:00 a.m. that day, OIM Jimmy Harrell had advised the rig crew to conduct the negative test via the drill pipe. After Jimmy finished his discussion I spoke and advised the rig crew members that BP engineering in Houston had instructed the BP well site leaders to conduct the negative test via the kill line, so I advised the rig crew and supervisors that BP preferred we conduct the negative test via the kill line. No one at the meeting made comments or objected to a kill line negative test and therefore when I arrived on the rig floor that evening I was somewhat surprised the rig crew and senior toolpusher had arranged the negative test to be conducted on the drill pipe.

The toolpusher and I had a short constructive discussion about the piping arrangement and he advised me that he and the rig crew were familiar with conducting negative tests via the drill pipe and not via the kill line. So, after a little more discussion I said, "Okay, let's test via the drill pipe."

The rig crew and supervisors agreed that the SOBM to seawater displacement went according to plan—a fact that showed up in the BP Investigation Report.[3] After the displacement, the drill pipe, which was run in the well to 8,367 feet, was full of seawater up to the rig floor.

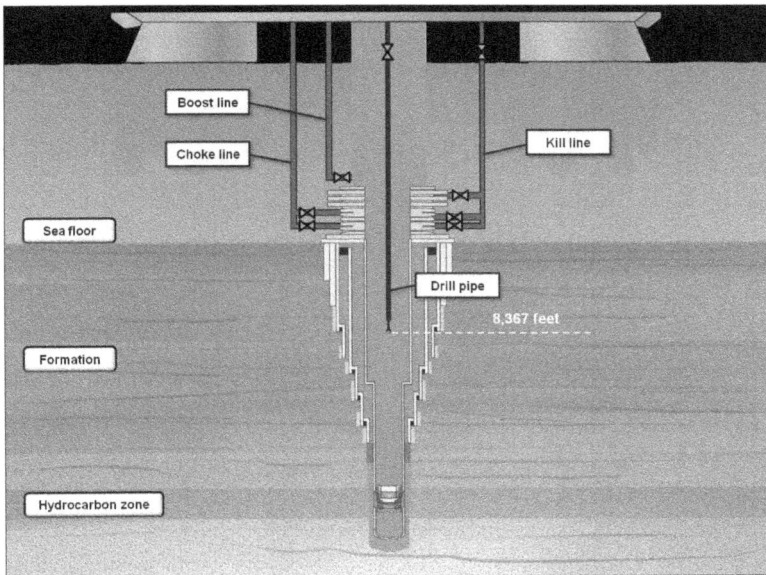

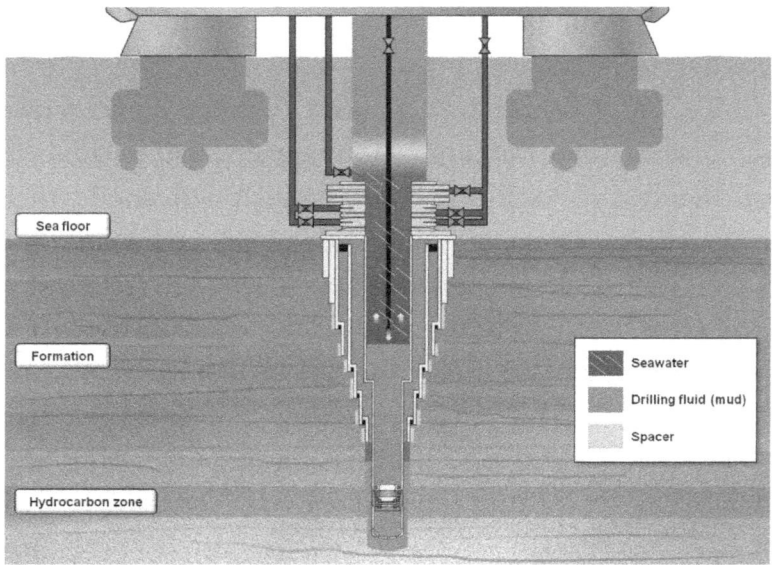

When the mud pumps were shut-down at the completion of the first displacement there was 1,262 psi final shut-in pressure on the drill pipe. That's normal when conducting this type of negative test.

The well pressure—pressure consisting of a combination of hydrostatic pressure of seawater in the drill pipe (8.6 pounds per gallon) plus hydrostatic pressure of the weight of the mud in the production casing below the drill pipe (14 pounds per gallon), was underbalanced. That means the pressure in the well after displacement was less than the known Macondo formation pressures.

Very simplistically: Imagine squeezing a plastic straw filled with mud and then squeezing a straw filled with water. There would be a big difference in pressure within the straw, although the straw filled with water doesn't necessarily break.

To start the first negative test, the driller called via walkie talkie to the Halliburton cementer standing by at his equipment below the rig floor. He notified the cementer to bleed-off the drill pipe pressure and call via walkie talkie when the cementer noted a "no flow" and then report how many barrels of fluid were bled-off. In less than sixty seconds, the driller announced in the driller's cabin, "We have a no flow." Fifteen barrels of fluid were bled off, which is equal to 630 gallons. There were no surprises.[4]

The driller announced the "no flow" at 5:27 p.m. The drill pipe remained open from 5:27 until 5:32 p.m., that is, for five minutes. If there was any communication with Macondo reservoir pressures leaking into the Macondo well, there would have been immediate, massive and continuous flow out of the open drill pipe. The cased well and drill pipe was underbalanced to Macondo formation pressures from 865 psi to 1,996 psi, which is a very large differential underbalance in a well and drill pipe. There was no flow from the drill pipe. The first negative test was successful and irrefutably established well integrity.

While the drill pipe was open, the senior toolpusher and I continued our conversation about negative testing on the drill pipe versus on the kill line. The two tests would be the same, we figured, so let's leave the test on the drill pipe. I told him I was going to leave the rig floor, go downstairs, and discuss the proposal with the other BP well site leader. Before I left the rig floor, we conducted a brief flow check of the kill line and there was no flow from the kill line.[5]

At 5:32 p.m., I asked the toolpusher to close-in the negative test on drill pipe. I was going to leave the rig floor to have a discussion with the more senior BP well site leader to discuss if we could leave the negative test on the drill pipe rather than on the kill line. The senior toolpusher requested we simply leave the drill pipe open for twenty to twenty-five minutes longer and conduct the negative test on the already arranged and open drill pipe. I said I'd make the recommendation, so I left the rig floor.

A common-sense test should be discussed here. On April 20, 2010 I had more than thirty years of diverse oil and gas drilling experience and I would never leave a rig floor if there was any type of well trouble. Neither would any other professional I know.

At approximately 6 p.m. that same evening, the senior daylight toolpusher left the rig floor to attend a meeting and later he went to bed. Someone with his decades of experience would not leave a rig floor to go to bed if there was any type of trouble in the well. Unfortunately, no one conducted common sense analysis in the entire course of the *Deepwater Horizon* investigation. Never.

I walked down to the well site leader's office and informed Don Vidrine that we had just conducted a successful five-minute negative test on the drill pipe because the rig crew and the daylight senior toolpusher preferred to conduct the negative test via the drill pipe.[6] Don said that the Houston engineers required that we negative test via the kill line, so Don politely said, "Bob just go back to the rig floor and tell the guys to switch over to the kill line." I said, "Okay," and headed back up to the driller's cabin.

The Infamous Tests: Part II

I arrived back to the driller's cabin on the rig floor at 6 p.m. after a discussion in the well site leader's office. The daylight toolpusher was having a change-over discussion with the night toolpusher, Jason Anderson, who had just come on duty. Rig crew members changed work tours at 11:30 a.m. and 11:30 p.m., however, toolpushers and BP well site leaders changed tours at 6 a.m. and 6 p.m.

I stood away from the two for a couple of minutes as they spoke; they finished and looked in my direction. I told them that BP engineers required that we test via the kill line.

The night toolpusher spoke up immediately and explained that he had tested via the kill line several times. He explained two different test procedures he had used in the past.

During my conversation with him, the daylight toolpusher left the driller's cabin.

Immediately after I had a conversation with the night toolpusher, Don Vidrine arrived on the rig floor and spoke with him concerning the best method to conduct a negative test via the kill line.

While Jason and Don were talking I noticed a slow pressure increase in the drill pipe. When Don Vidrine and Jason finished their conversation, I pointed out the slow pressure increase in the drill pipe to Jason. He immediately stated that the drill pipe pressure increase was common when using the BOP annular preventer rubber during negative testing. Tremendous pressure is exerted on top of the annular rubber after the circulation. Imagine pressing down on a beach ball with your hand and how that would temporarily deform it. With pressure as high as 1,400 psi and weight on top of the deformable rubber as high as 400,000 pounds, the annular preventer rubber would deform. That deformation exerts a pressure below the annular rubber—which causes the shut-in drill pipe pressure to increase.

I asked Jason how high the pressure in the drill pipe would increase and he said the pressure will increase until it reaches the differential pressure exerted on top of the annular rubber. It bears repeating that Jason had been onboard *Deepwater Horizon* since it launched in 2001 and his experience was indisputable.

Whipping out my calculator, I determined that his knowledge suggested the maximum pressure should be 1,404 psi.

Jason was comfortable with what was happening. He stated the pressure in the drill pipe would increase to approximately the differential pressure of 1,400 psi, and then stop.

I was skeptical of the explanation, but I had never used a deformable annular rubber before to conduct a negative test. An experienced toolpusher is a reliable source for information on a drilling rig and Jason sounded completely credible and confident in the explanation. He referred to this as the bladder effect.

I left the driller's cabin at approximately 6:20 p.m. and returned about thirty minutes later. When I returned I noticed the drill pipe pressure had increased and stopped at approximately 1,400 psi, exactly as calculated and predicted. Jason Anderson and the rig crew were arranging piping on the rig floor to conduct the second negative test via the kill line and routing any flow returns into the rig mini trip tank. (The trip tank is a small, calibrated tank below the rig floor. It's not fully enclosed so pressure never builds in the tank, but it has electronic calibration installed to measure volume and the volume measurement can be observed in the driller's cabin.)

When Jason returned to the driller's cabin he explained to Don that he had arranged piping from the kill line to the mini trip tank and the rig crew was ready to conduct the negative test. I asked if there were a gauge that could be observed if there is flow into the mini trip tank. Jason said, "Yes," so I found the gauge and volunteered to sound-out if volume increased.

Jason instructed rig crew members to open two valves to the mini trip tank and within seconds there was an increase of volume into the mini trip tank; the volume increased from .6 bbls to .8 bbls (eight to nine gallons) in approximately twenty seconds and then all flow stopped. The kill line remained open for the next thirty minutes from 7:20 to 7:50 p.m. Rig supervisors monitored for flow for thirty minutes; there was no flow. The second negative test was deemed successful—a fact asserted by OIM Jimmy Harrell in testimony from what he had been told by the toolpusher on site.[7] I left the driller's cabin at 7:50 p.m., an hour and fifty minutes after my tour ended, and went to the galley and ate a bowl of soup. After that, I went to my bedroom and fell asleep.

Science and Common Sense

The negative tests quickly became infamous as BP scrambled to give the public an explanation for the blowout. In not just the days after the tragedy, but even years after it, science and common sense were submerged like so much of the physical evidence related to *Deepwater Horizon*.

Fake versus Real Science

BP was being pressed to give a reason for the *Deepwater Horizon* catastrophe. Brushing aside any inconvenient science that confirmed the negative tests were good, some BP engineers speculated that an influx of oil and gas had started to come up the well and that is what caused the 1,400-psi drill pipe pressure. That assertion made it into the BP Investigation Report about the accident that was released in September 2010.

They ignored the fact that there is a formula to calculate gas migration of a shut-in well. Gas should migrate up at 250 feet/hour. If you get an influx, it continues to migrate up at that rate. The pressure is shut in—it has nowhere to go but up. And because the influx is shut in, it has no room to expand. BP's assertion therefore makes no sense. If the pressure were from the well, it wouldn't stop migrating suddenly and level off. Flow from the reservoir would keep migrating upward toward a rigid barrier and the pressure would continue to rise. It didn't. Opening the kill line would have fluids pushing upward, flowing freely. There was no flow.

BP asserted that the kill line must have been plugged; that's why there was no flow. But in the post-mortem, they found that it was not plugged. That finding confirmed that the 1,400 reading we saw could not have been reservoir pressure.

The drill pipe pressure was thoroughly and carefully discussed among the rig supervisors prior to beginning the second negative test.

In summary, even without Jason's bladder-effect explanation, science is not on the side of BP regarding its conclusions about the negative tests. First, the drill pipe pressure increased steadily from 200 psi to 1,400 psi in thirty minutes, and then once pressure reached 1,400 psi it stopped suddenly. An influx of oil and gas in a 13,000-foot vertical cased well would not suddenly stop migrating in the middle of the well! And again, if the source of 1,400 psi drill pipe pressure was from a Macondo reservoir pressure and both the kill line and drill pipe were tied into the same cased well, which they were, then the kill line must flow when it was opened during the second negative test. As the experienced people around me on the rig saw, there was no flow from the kill line for 30 minutes. BP and Transocean supervisors determined the second negative test successful due to one single criterion, the observation of no flow for 30 minutes.

Unfortunately for Don Vidrine and me, BP's flawed conclusions formed the basis of indictments on eleven cases of manslaughter that were pinned on us.

The Foundations of Fake Science

BP officials had neither careful engineering nor careful risk assessment at the core of their decision making in those final days of *Deepwater Horizon*'s operation. There had been problems with the well that put the Macondo effort behind schedule by more than a month and that translated to being at least $40 million over budget. Changes to temporary well abandonment procedures were part of the last-ditch effort to cut losses.

A big problem with Macondo stemmed from the drilling team encountering unexpected formation pressures. They were forced to stop drilling at 18,360 feet due to hole problems created after encountering *inverted formation pressures* in the well from depths 17,684 feet to 18,223 feet. "Inverted" simply means that formation pressure in the upper formations were higher than pressures in the lower formations; normally, pressures increase in a well as a well gets deeper.

Due to inverted formation pressure, the weight of the drilling fluid in the well must be raised to keep and maintain Primary Well Control (hydrostatic weight of the fluid in the well greater than formation pressure), but due to the higher mud weights when the drill bit encounters the lower formation pressures deeper in the well, the mud can break down the weaker formation rock and flow into the formations; that is called "lost circulation." Lost circulation was an expensive problem in the Macondo well and because the well was already so much over budget, the Macondo team in Houston decided to temporarily abandon the well.

The inverted formation pressure problems forced BP engineers and Halliburton engineers to consider risky cementing strategies when planning the final production casing cement job. Cementing contractor Halliburton's recommendations were disregarded and BP ordered the Transocean rig crew to take a shortcut to reduce project time.

Halliburton normally recommends that a well is circulated to at least "bottoms up" before pumping cement.

This involves circulating all the fluid and possible gas from the bottom of the well all of the way to the surface and into the mud pits; the area from the bottom of the well to the surface is called "bottoms up." By circulating in this manner, the mud ahead of the cement will likely have a uniform weight during the cementing process, and reservoir gas that would have accumulated at the bottom of the well will be circulated out of the well. Any spotty weighted mud, which often contains drilling debris, can also be circulated out of the well. When drilling debris and reservoir gas is circulated out of the well prior to cementing, the mud column ahead of the cement is uniform, among other advantages.

To achieve bottoms up on the Macondo well 2,760 barrels of fluid had to be circulated prior to starting the production casing cement job. At 4 barrels per minute, circulating bottoms up could take more than eleven hours. The BP Macondo drilling team chose to save approximately twelve hours of rig operation time and chose to circulate only 111 barrels prior to starting the production casing cement job.

Because the Macondo drilling team chose not to circulate bottoms up, there was reservoir gas on the annulus side of the production casing and likely some debris from the well on the annulus side, and both gas and debris in the annulus could affect the accurate observation of some critical pressures at the end of the cement job. The Macondo drilling team was aware of these factors prior to cementing, but chose not to circulate bottoms up, regardless of how it might affect the placement of the cement.

The General Counsel's report referenced earlier does not hold back in criticizing BP and Transocean decision makers for brushing aside good science and technical expertise as they drove ahead with plans that sacrificed safety:

> Despite making multiple changes over the last nine days before the blowout, the Macondo team did not formally analyze the risks that its temporary abandonment procedures created. The Macondo team never asked BP experts such as subsea wells team leader Merrick Kelley about the wisdom of setting a surface cement plug 3,000 feet below the mudline to accommodate setting the lockdown sleeve or displacing 8,300 feet of mud with seawater without first installing additional physical barriers. It never provided rig personnel a list of potential risks associated with the plan or instructions for mitigating those risks.[8]

The Value of Common Sense

All drilling rig personnel have a right and an obligation at any time to "stop the job" if they detect anything unsafe. This stop-the-job policy has been implemented on all BP drilling rigs and other oil and gas production facilities for many years and most rig personnel practice that obligation. "Stop the job" is reinforced on BP contracted drilling rigs and will always be positively received by all management and supervisory personnel. No employee can ever get in trouble by stopping a job!

While I was on the rig floor conducting or observing the two negative tests, no Transocean employee or any other employee voiced any concern about the safety of the negative testing. Many workers on *Deepwater Horizon* had been employed on that rig for many years and those long-term employees had assisted in conducting negative tests using the same procedures followed on April 20 for many years.

As I also noted above, no rig supervisor would leave the drilling rig floor if that supervisor detected a well problem. The senior toolpusher left the rig floor after the first successful negative test and was in bed reading a book at the time of the blowout. OIM Jimmy Harrell came through the driller's cabin sometime before 6 p.m., spoke with the driller and toolpusher, and left the rig floor. Jimmy was in the shower preparing to go to bed at the time of the blowout and I was asleep.

It seemed that no so called "expert"— none of the Macondo investigation teams and certainly none of the lawyers and investigators at the US Department of Justice—considered simple common sense when they assessed what happened on *Deepwater Horizon* on April 20, 2010.

CHAPTER FOUR

36 HOURS AND THE WEEKS AFTER

had only hours left to go on the rig when the explosion occurred. I was in my tiny room on *Deepwater Horizon* and had already thrown my boxer shorts, socks and pants into an overnight bag that would leave with me on my helicopter flight to Houma, Louisiana.

Walter Pavlo, an author who specializes in white collar crime, succinctly captured my first hour after the blowout in an article for *Forbes*:

> Around 10:00 p.m., Kaluza awoke to a jolting fire alarm that brought him from a deep sleep to standing beside his bunk in seconds. There was a booming announcement over the platform's loud speaker, "This is not a drill!" At that moment he did not know about, nor had he heard, an explosion that occurred three stories above his cabin on the main deck that was responsible for the deaths of eleven rig workers, which led to the largest oil spill in U.S. history.

Kaluza's room was completely dark and after fumbling for the switch discovered that it did not work. He emerged from his room, walked down a corridor to a watertight door, opened it, went up two stories of spiral staircase and opened another. Smoke rushed in and he snapped it back shut. He walked the opposite direction down the corridor and saw a person with a flashlight. Kaluza introduced himself by saying, "I'm the new guy on the rig and can't find the passage to the lifeboats." Kaluza followed the stranger to the lifeboats and his passage to safety.

When the lifeboat was a few hundred yards from the rig, five minutes after dropping into the water, the operator opened the door and Kaluza saw the yellow flames engulfing the *Deepwater Horizon*. He knew that the situation was dire, but what he did not know was that he would be blamed for it.[1]

Anyone on an offshore drilling rig cannot sleep through the sound of a fire alarm. The noise is deafening. The alarms go off inadvertently from time to time because there are sensors all over the rig and it's possible to trip the sensor without meaning to. Regardless of the reason it goes off or when it happens, everyone on board must rally for a fire alarm. You might be in a deep sleep, but within minutes you are fully clothed and at your muster station. It's just part of living on an offshore rig.

I heard running in the hallway.

The first announcement was, "Go to your muster station." I realized something might really be wrong when I pushed the button to turn a light on at my bed and nothing happened. I had only slept in that room three nights and wasn't sure where everything was, so I cracked open the door of my room, thinking that emergency lights would be on in the hall. Nothing. I was standing in a pitch-black room groping for the socks and work pants I knew were nearby.

The sound of the announcer's voice now had obvious anxiety: "This is not a drill! This is not a drill! Everyone to your muster stations."

When I went to the stairway and couldn't access it, I knew something serious had happened. Ceiling tiles had fallen on the stairway and blocked it. I knew only one way up and this was it.

The next announcement was, "A beam has fallen into the galley. Do not muster to the galley." That was my muster station.

As Pavlo's article states, the reason I was able to go anywhere and ultimately find my way to safety was that a stranger with a flashlight guided me to the pipe deck and I could navigate from there.

The New York Times opened its December 25, 2010 article on the disaster with a riveting and accurate description of what we faced:

Crew members were cut down by shrapnel, hurled across rooms and buried under smoking wreckage. Some were swallowed by fireballs that raced through the oil rig's shattered interior. Dazed and battered survivors, half-naked and dripping in highly combustible gas, crawled inch by inch in pitch darkness, willing themselves to the lifeboat deck.[2]

As I got to the lifeboat, a man behind me screamed, "The derrick's burnin' up! It's gonna fall!" He seemed frozen in terror.

"You better get in this boat!" I yelled to him.

"I'm not getting' in that boat!" He said he was claustrophobic. The lifeboat had a deck with a sheltered area; with fire and smoke all around, the sheltered area was where we needed to go.

I'm not one for hyperbole or drama. "This is probably your best option," I told him.

Those of us who were uninjured helped others onto the lifeboats that could hold about seventy-five people. Not everyone went onto the lifeboats, though. Some jumped into the water; some took rafts. I don't know the stories of the people who jumped other than from reading about them.

My guess is that about 50 percent of the people I encountered were zombie-like, looking as though they were stunned and in shock. They had to be herded, coaxed, and nudged into the lifeboats.

One man in a managerial role tried to take a roll on the life boat deck, but I told him that would be impossible. People were wandering around in panic and abject disbelief, overwhelmed by shock and surprise. I recommended that we get everyone inside the boat before attempting to take a roll.

We were getting people inside and someone pushed me toward the door, "You better get in!" Just before I got in, someone scooted ahead of me. It was the guy who said he was claustrophobic. We waited until no stragglers were in sight on the rig, and then waited a little longer. Soon, we dropped smoothly into the water about sixty feet below. I heard the gear kick in and we started to motor away from *Deepwater Horizon*.

When we reached the offshore supply vessel *Damon Bankston* that had navigated over to rescue us, they dropped a rope ladder with about eight wooden rungs. Our boat was heaving up and down so some of us stood and helped people mount the ladder to stop them from getting pinched by the *Bankston*.

Once the people from our boat were on the *Bankston*, their crew put up two card tables, sat down, and a couple of people started taking names. Everyone was expected to check in, although some could not do it personally.

With our initial panic burned off, the *Horizon* crew were now in various states of shock. Some were injured badly enough that Coast Guard helicopters had to extract them from the deck and take them to shore. We saw a lot of heroism that night.

I looked for Don Vidrine, who ended up being on the other lifeboat. The time lapse was under thirty minutes between the time the first and second boats arrived at the *Bankston*. When he disembarked, I said, "Don, what the hell happened?"

"I don't know, Bob. I don't know."

I've been asked if anyone on board had a cell phone and took pictures or recorded thoughts. All I can say is this: When it comes to fire, any delay of escape is stupid. I once knew a guy who went back into a frat house during a fire to try to save his girlfriend and he died, too. When your profession is to do a job on an offshore drilling rig—arguably one of the most dangerous places to work, even though there are redundant safety systems galore— you know you waste no time in rushing to safety in a catastrophic event. Taking photos is not for people like us in a life-and-death situation.

We learned from each other that, one or two levels below the explosion in the office areas and living quarters, people had been slammed against walls. The rig's rugged construction, with everything reinforced and built to withstand mindboggling punishment, fell victim to the violence of the explosion.

From the *Bankston*, parked all night at a safe distance from the rig, we watched the flames change color on *Deepwater Horizon*. It was a dark color at first because the fire was burning oil. Later it turned to orange, probably due to chemicals in the gas, and then white—the look of pure gas burning. The dark, diesel-smoke appearance was short-lived. Then for a while, more than an hour at least, we saw the yellow-orange glow. After that, it went to white. It was too bright to look at directly at night.

I flashed back to a farm fire near the University of North Dakota where I did my undergraduate work. A couple of us saw a flame in the distance so we drove toward it. A farm house was on fire. Two cars near the house got incinerated. The fire was so hot that, even at the road about three-hundred yards from the house, we had to back up our car. The heat drove us back and back and back. A fire is not an invitation.

Bankston sat in the water far enough away from Deepwater Horizon so that those of us on board could not feel the heat. I don't even know if we were half a mile or a mile away. The water was dark, we were safe, and that's all we knew. The mood was somber as we watched the Coast Guard evacuate severely injured people from the deck.

Not long after that, on April 21, 2010, I was one of the people interviewed by a hastily assembled team trying to start putting the pieces of the tragedy together. I was interviewed by two people from MMS—Investigator Glynn Breaux and Specialist Randy Josey—and one from the United States Coast Guard, Lieutenant Angelique Flood. Her colleague, Lieutenant Barbara Wilk, was present, but didn't ask any questions. Another person in the room who did not question me was Cliffe Laborde, an attorney for Tidewater, Inc, which owned the *Damon Bankston*. I didn't know it at the time, but the interview was recorded.

Later, after I faced criminal charges and my attorneys were assembling evidence, they were thrilled at having a recording of this interview. I had rattled off everything I'd observed and knew, and it not only made engineering sense, but was also delivered without any planning or preparation. Investigators want to dive into witness stories as soon as possible after an event to get raw information, untainted by conversations with other witnesses or advice from anyone on what to say or not to say.

Early in the conversation, Breaux took the lead in asking technical questions about the tests we'd been doing aboard the rig. He had already spoken with the surviving senior toolpusher I'd worked with on the negative tests, so he was acquainted with some of the events of April 20. He stepped through a few of parts of the tests and asked what happened after that. Here is what I replied:

INV. BREAUX: Well, at this point, it sounds like you did it both ways, and that way you were covered. . ..

MR. KALUZA: We may have had that drill pipe open for 10 minutes at the most, and it bled to zero, so it was [good]. See, that's the thing about this whole situation. I mean, you bleed the drill pipe down, and the drill pipe is down to 8367 feet with seawater or maybe a little intermingling of the heavier mud in some places pretty much until you get it all pure seawater. But the drill pipe is down to 8367 feet, and when we bleed it down, then it's a long stem with seawater in it. And when you bleed it and you open it, and nothing flows, you're saying to yourself, you know, this is a type of negative test that's showing that we don't have anything coming from the casing, nothing that's coming out of the casing and into the pipe. But then we did like I said, then we did shut the drill pipe in and convert over to the kill line, and that's how we monitored the thing. That was how.[3]

Here is another excerpt from that transcript:

MR. KALUZA (cont): So we bled it off and got 0.2 barrels, ah, and then it stopped. So now we're doing a negative test on the kill line, right, and we monitored the mini trip tank. The mini trip tank stayed steady at 0.8 barrels for the entire 30 minutes, and I stayed on the rig floor. And that lasted until 7:50 p.m. At that time, I mean, I was going to relieve Don early, because I was going in today, and he was switching over to days. So the way we do it is I relieve him about two in the morning, and then he goes to bed, takes a nap for two hours and then comes and relieves me at six when I'm getting ready to get on the helicopter. So I wanted to stay to see if we got the successful negative test, so we got a successful negative test, no flow whatsoever, monitoring from the kill line in the BOP stack up to the surface. With seawater, it's all the way down to 8,367 feet. That is equivalent of about 2340 psi. We had to calculate it (inaudible), but it was about 2340 psi is the equivalent differential pressure right there. So at 7:50, the test was over, and I said to Don, "I'll see you at two a.m." And I went to bed.[4]

We returned to shore around 3 a.m. Family members crowded around, looking for their loved ones.

We were taken to one of the office buildings at the dock. After providing a urine sample, we were sent to speak with an attorney about what we could and couldn't say about the event. BP had arranged for someone to drive me to Houma. Someone gave me $200 for incidental expenses, and then I flew home to Nevada, but had a short layover in Denver. I called my mother who lived there and told her I was fine and would call her later when I got home. Someone from BP had already let her know that I had survived the explosion.

In a few days, I got a call from BP's human resources department asking how I was doing. Right after the accident, people like this from BP who contacted me sounded very concerned. They could only imagine what those of us who had survived had gone through and their compassion leaked into every conversation. I even got a handwritten note and card from one of the executives whom I'd met before.

Less than a month after the incident, I got another couple of important calls—one from HR telling me I'd been put on paid administrative leave and another from a BP attorney telling me I had been assigned an attorney. He gave me the number for Shaun Clarke, who ultimately worked with his partner David Gerger, and their colleagues Dane Ball and David Isaak, to restore my freedom, my reputation, and my quality of my life.

CHAPTER FIVE

12 MORE SUSPECTS

Peoplε who have direct ties to the cause of the disaster did not work on the rig. Their offices in various parts of the world kept them, for the most part, at physical distance from the tragedy—a fact that hints at the mental distance they had from the needs onboard *Deepwater Horizon*.

The Onshore Players

It is not possible for me to say with certainty that every person mentioned in this chapter knew that *Deepwater Horizon*'s BOP equipment was out of compliance in advance of the blowout. Starting with the email thread included in the Introduction, however, it is clear that many of them did. It's also an undisputable assertion that the cozy relationships that BP and Transocean officials had with rig auditors and federal employees contributed to lax (or zero) enforcement of standards, policies, and federal regulations.

Macondo Operations and Engineering Team

Until March of 2010, one month before the Macondo blowout, Ian Little was the Macondo Wells Manager.

Little managed John Guide, Wells Team Leader, and David Sims, Engineering Team Leader. Guide managed senior operations drilling engineer Mark Hafle and operations drilling engineer Brian Morel, as well as four well site leaders that worked on Deepwater Horizon; he also managed operations drilling engineer Brett Cocales. As noted in the Introduction, if they received an email about the results of the audit, then they knew the BOP equipment was out of federal regulatory compliance for both maintenance and inspection.

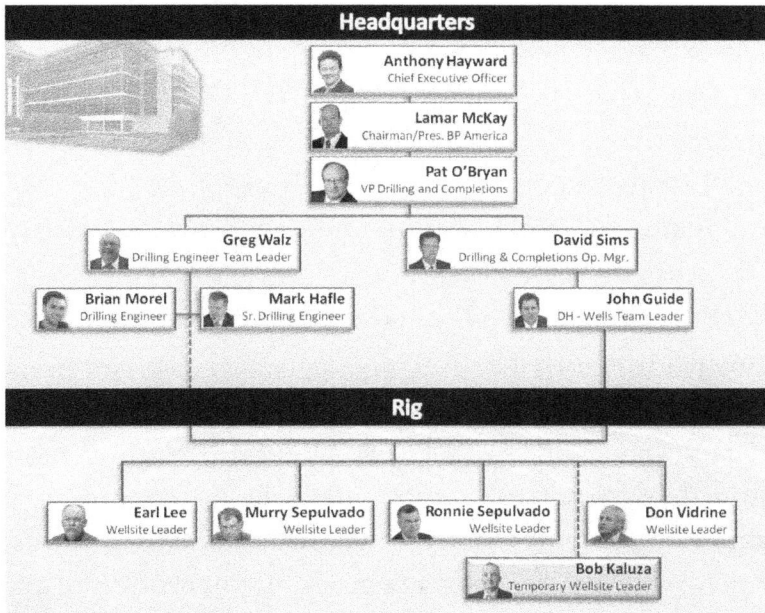

In March 2010, one month before the Macondo blowout, the Macondo wells management and engineering team changed. Ian Little left the team and David Sims became the new team leader. Guide remained Wells Team Leader reporting to Sims, and Walz was promoted into Sims' old position. Walz reported to Sims; Hafle and engineer Morel also reported to Sims first, then to Greg Walz; Cocales continued reporting to Guide first.

In terms of accountability for BOP compliance with the US Code of Federal Regulations, the senior person on the chart who could have shut down *Deepwater Horizon* operations to ensure compliance was Pat O'Bryan. O'Bryan was short-lived in his position, assuming his VP role in 2010 before the blowout—he was actually on *Deepwater Horizon* April 20—and exiting just after his testimony at the August 2010 joint hearings by the US Coast Guard and the Bureau of Ocean Energy Management, Regulation and Enforcement investigating the causes of the explosion. An interesting aside is that his LinkedIn profile incorrectly lists when he served as VP, putting the "less than a year" term as during 2011.[1] I suppose that's one way to distance yourself from disaster.

O'Bryan's predecessor, Kevin Lacy, was there much longer than he was. As VP of Drilling & Completions from July 2006 to December 2009, he theoretically had ample opportunity to ensure that the BOPs complied with federal regulations. And since the BOPs officially were out of compliance as of April 2006, BP had been in violation of federal regulations throughout his tenure.

In 2008, 2009 and until March 2010, Ian Little was accountable[2] for ensuring his drilling team adhered to BP Policy and policy of the Minerals Management Service (MMS), a division within the Department of the Interior. By BP policy, a person labeled accountable "accepts accountability for decisions" and "there can be only one accountable party." After March 2010, when Little left, the position was re-titled BP Drilling and Completions Operations Manager and given to David Sims. He then became the accountable party for BP Policy adherence and MMS Policy adherence on the Macondo drilling team. Both had the authority to shut down rig operations immediately and order the BOP inspection.

In short, BP's RACI chart (R=Responsible, A=Accountable, C=Consulted, I=Informed) essentially exonerates anyone above or below Little of culpability for the BOPs' non-compliance through March 2010, a month prior to the blowout, since "there can be only one accountable party." And after that, the sole person accountable was David Sims.

To complicate matters further, the VP preceding Ian Little should have ordered a BOP inspection even before Little assumed his job. By the time Little took over, the BOP inspection was already overdue by two years.

In the disaster scenario, John Guide is the other critical person belonging to the BP executive team—and he absolutely knew the BOPs were out of compliance as the email thread provided in the Introduction documents. To reiterate his response to auditor Norman Wong's alert about the problems: "Norman throwing DW Horizon under the bus."[3]

Guide was the Wells Team Leader at the time of the blowout, so he led the engineering and operations teams of the Macondo well. Guide likely did not have the full authority to stop drilling and shut down *Deepwater Horizon* operations for a BOP inspection. Nonetheless, he had oversight responsibilities as well as a moral obligation to inform Little or Sims that the *Deepwater Horizon* BOP equipment was not in compliance with API regulations as well as the Code of Federal Regulations—and to demand immediate compliance when he learned of the problems from Norman Wong in October 2009.

Every member of every BP drilling team worldwide was required to be knowledgeable of BP minimum drilling standards described in *Drilling and Wells Operations Practices* (DWOP) and to comply with all defined practices at all times. In everyday terms, drilling managers, engineers and supervisors at BP had a careful, well-thought-out and proven drilling standards system that worked. If drilling managers, engineers and supervisors followed minimum standards, a drilling "train wreck" should never occur, certainly a well blowout catastrophe should never occur.

If any one of these people at BP in the upper section of the chart had simply followed BP Corporate Safety Management Policies, there would not have been a Macondo blowout, loss of life, and oil spill.

Two people who also belong to this scenario are Harry Thierens and Jonathan Sprague, both of whom received the audit report from Norman Wong. Clearly, Wong was becoming frustrated with the lack of response from John Guide. His October 6, 2009 email to Theirens and Sprague began:

> As Wells Director and Wells Engineering Authority, I just wanted to bring to your attention the most significant findings from the recent rig audit of the Deepwater Horizon. You may have been made aware of this already but just in case you were not, thought it best you should know.[4]

Theirens, BP Executive Vice President for Drilling & Completions, later testified under oath that the reason the BOP failed was that it was connected incorrectly.[5] Not a word came out of his mouth about non-compliance with federal regulations.

Sprague, a senior vice president who had managed BP's Gulf of Mexico operations, was more central to the discussion of Kurt Mix's prosecution than the BOP discussion. Ultimately vindicated, Mix was a BP engineer accused of obstructing justice in the Macondo case by deleting text messages, many of which supposedly were between him and Sprague, his supervisor. Mix was not onboard *Deepwater Horizon* at the time of blowout; rather, he was brought in to help stop the oil spill. Kurt Mix was not scapegoated as Don Vidrine, Dave Rainey and I were, but he was falsely accused of trying to obstruct justice as part of an ongoing effort to make anyone except the accountable parties at headquarters look like criminals.

And then there were those of us assigned to the rig. Don Vidrine and I were serving as well site leaders (*aka* rig supervisors) for BP at the time of the blowout. My job, scheduled to run from the morning of April 16 to the morning of April 21, was to work with Don to conduct positive and negative tests that preceded abandonment of the Macondo well. As I explored in Chapter Three, one of those tests ultimately was what BP asserted was the inciting incident in the blowout, hence, the subsequent indictments of us on eleven counts of manslaughter.

Macondo Operations and Drilling Supervisors

Of course, BP executives and managers weren't the only off-the-rig people able to force action to bring equipment into compliance with federal regulations. There was also a Transocean team of decision makers.

Senior managers at Transocean, the company that owned *Deepwater Horizon*, are as culpable and accountable for the Macondo blowout as senior managers at BP. Some of them at Transocean did voice serious concerns about the maintenance condition of the BOP equipment in emails to BP managers through numerous e-mails, but at no time did Transocean managers "stop the job" and compel compliance with Transocean's safety management system.

BOPs are a preventative maintenance schedule item and therefore drilling contractors are expected to stay ahead of problems by demanding that equipment be properly maintained. In 2009 and 2010, there were e-mail exchanges between managers at BP and Transocean discussing if and when BP was willing to stop drilling for BOP maintenance. BP managers informed Transocean managers they were delaying BOP maintenance again and again. In fact, John Guide stated to Transocean rig manager Paul Johnson that "BP accepts responsibility if both annulars were to fail and the stack had to be pulled to repair them."[6] Note well: Guide did not step forward to accept any responsibility after the Macondo blowout.

The Rig Auditors

Who is Det Norske Veritas? Long term followers of Macondo news, listeners of interviews with many so-called experts and even media people that wrote about Macondo events may have never heard of Det Norske Veritas (DNV), however, managers and rig auditors of DNV could have changed the course of history and prevented the blowout. To my knowledge, their identities did not become part of media coverage or public records related to the investigations or trials.

DNV is a Norwegian company headquartered in Oslo with a long history of expertise in technical inspection and evaluation of merchant vessels. In 2010, DNV was contracted by the Mobile Offshore Drilling Unit (MODU) Registration Flag State, the Republic of the Marshall Islands, to audit the *Deepwater Horizon* rig annually to ensure that Transocean's comprehensive safety management system (SMS) was in compliance with the International Management Safety Code (ISM).

The RMI [Republic of the Marshall Islands] and the USCG were mandated by international and U.S. regulatory requirements to perform inspections and examinations on the MODU [Mobile Offshore Drilling Unit]. The RMI did not physically evaluate the MODU. All of DEEPWATER HORIZON Ship Statutory Certification Services were performed by the recognized organizations (RO) acting on behalf of the RMI. ABS [American Bureau of Shipping] acted as the RO for the review and survey of technical issues such as engineering and design, while DNV was the RO for the review and audit of the safety management system (SMS) for compliance with the ISM Code.

RMI was responsible for conducting oversight of whether DEEPWATER HORIZON design, manning and operations were in accordance with international standards and flag state regulations. RMI delegated these duties to two recognized organizations, American Bureau of Shipping (ABS) and Det Norske Veritas (DNV). DNV was responsible for issuing ISM certificates, the Document of Compliance (DOC) and the Safety Management Certificate (SMC).[7]

International maritime ship and rig auditors like DNV, representing Flag States, conduct critically important and extensive annual audits of Mobile Offshore Drilling Units (MODUs) required by international law. Recognized organizations (RO), such as DNV, carefully evaluate and ensure that MODU Contractors' Safety Management System meets ISM compliance. If an RO auditor discovers that a MODU Contractor's SMS does not fully comply with ISM code, the auditor—in this case DNV—will refuse to issue the Document of Compliance (DOC) or Safety Management Certification (SMC) which would cause the MODU Contractor to shut-down the vessel or floating drilling rig until necessary changes had been made to comply with International Management Safety Code. A MODU Contractor would lose millions of revenue dollars if their floating drilling rig was shut-down due to non-compliance with ISM code.

The outcome of annual RO floating drilling rig audits is critical to the profitability of all international offshore drilling companies; that included the outcome of annual rig audits by DNV of the MODU *Deepwater Horizon*.

International RO auditors can take floating drilling rigs out of service for failure to comply with ISM codes.

DNV conducted annual audits of *Deepwater Horizon* to ensure that the Transocean Safety Management System met ISM code requirement. But since April 2006, Transocean's SMS did not meet compliance with ISM code: ISM code requires full compliance with mandatory rules and regulations which included full compliance with the US Code of Federal Regulations.

DNV conducted annual ISM compliance audits on *Deepwater Horizon* during 2006 through 2009 and after each audit failed to notify RMI or BP or Transocean that Transocean's SMS violated ISM safety code. Nonetheless, each year, DNV issued ISM certificates of compliance.

By not notifying the Maritime Registration Flag State, Republic of the Marshal Islands, BP or Transocean, DNV violated its duties as a "recognized organization." Additionally, in 2008 and 2009, DNV deviated from its own audit Recommended Practices concerning BOP inspections. According to that document (DNV-RP-E 101, October 2008):

It is DNV's recommendation that a major overhaul/ inspection of Blow Out Preventers and other well pressure control equipment used for Drilling, Completion, Workover and Well Intervention operations, should be performed at least every five years. The purpose of this inspection is to verify and document that the equipment condition and properties are within the specified acceptance criteria as well as the specified recognized codes and standards, thus ensuring that documentation of the condition of the equipment is available at all times.

I cannot say for certain why DNV managers and auditors failed to notify government registrars at the Republic of the Marshall Islands that Transocean's SMS did not comply with ISM code in the Gulf of Mexico, but I can speculate. I suspect DNV had an overly cozy working relationship with Transocean and/or with BP, somewhat like the BP-MMS cozy relationship, and therefore re-issued ISM certification every year.

Here is the ultimate irony that addresses how determined DNV was to issue a Document of Compliance in 2010 to Transocean. This is another excerpt from the US Coast Guard Investigation Report:

Despite Transocean's record of non-compliance with the ISM Code, DNV failed to connect the dots and endorsed Transocean's DOC in Houston, Texas on April 21, 2010, at the same time that DEEPWATER HORIZON was engulfed in flames a couple hundred miles away in the Gulf of Mexico.[8]

Federal Employees

Chapter One's coverage of the comfortable relationship between MMS employees and certain representatives of BP and Transocean pinpoints transgressions and identifies individuals. The only point to emphasize here is that federal employees who were derelict in their duties bear some direct responsibility for the blowout. People like Frank Patton, the district drilling engineer with the MMS permits section who rubberstamped requests related to the well, helped build the time bomb.

CHAPTER SIX

5 BIG LIES

This chapter spotlights five big lies that drove media coverage, public opinion, and even legal proceedings in the *Deepwater Horizon* disaster.

Upfront, I will say that we need media to speak truth to power, a phrase defined powerfully by law professor Judith Sherwin as "a powerful nonviolent challenge to injustice and unbridled totalitarian forces, often perpetuated by government, sometimes not."[1] My great disappointment, which is shared by anyone who respects the mission and responsibility of news media, is that many reporters did not, would not, or could not speak truth to power. One result is the perpetuation of some of BP's and Transocean's lies about *Deepwater Horizon*; they still creep into articles.

When the blowout occurred, there were numerous national news reporters on the Gulf of Mexico beaches reporting daily, interviewing so-called experts about what caused the blowout and who was responsible. Few of these supposed experts were much more than talking heads. Some gave false information out of ignorance and some did it deliberately. When they did it deliberately, they played on the sentiments of millions of Americans who wanted to blame some human being for the blowout and oil spill, especially if that human being was a BP employee. When I left the courthouse on a warm evening on February 25, 2016 after being found not guilty, there wasn't one national news camera team outside the courthouse. In fact, after the first day of Defense witnesses, there were only one or two reporters who remained to hear the experts my legal team had called.

No reporter contacted me after the trial to understand why it took less than two hours for a twelve-person jury to determine a "not guilty" verdict of a BP employee in a case that made national news for years. Instead of accepting that justice had been done that day, and learning from the airtight evidence exonerating me, some reporters continue to file outrageous and false reports like the following:

The 2010 Deepwater Horizon explosion the largest oil spill in American history dumped an estimated 4 million barrels of oil into the Gulf of Mexico and left 11 workers dead. Initially, the Justice Department pinned the blame on two well site leaders Robert Kaluza and Donald Vidrine for not notifying engineers that pressure tests showed the well they were drilling was insecure. Unaware of the failed tests, workers continued to drill, leading to the ensuing well blowout and explosion. Kaluza and Vidrine were each charged with 11 counts of felony seaman's manslaughter and 11 counts of involuntary manslaughter, as well as violating the Clean Water Act. The manslaughter charges carried maximums of 8 to 10 years in prison, but neither man would serve a day.[2]

My editorial response to that paragraph of erroneous information and innuendo—something that came 14 months *after* my not-guilty verdict—is not suitable for print. States News Service should get a "guilty" verdict for shoddy reporting.

What I learned while being forced to endure American injustice is this:

- Politics can, and often does, supersede justice in America;
- A well-conceived Code of Conduct communicated by a corporate CEO can become meaningless when the business could lose money if everyone abides by it; and
- Most people, even ethical people such as professional journalists, can both inadvertently and deliberately become complicit in injustice when they allow their politics to flavor their ethics.

This chapter is about moral challenge. What types of unethical steps will individuals or in this case, corporations take to protect senior managers and save money?

Before I delve into five big lies that fueled public interest in putting an orange jumpsuit on me, take a minute to consider the motives for deceit and rationalization for cheating in some manner. Ironically, it may help you sympathize with the morally weak human beings who offered me up as a scapegoat.

As the authors of the bestseller *How to Spot a Liar* note, people lie because of love, hate, or greed.[3] They assert, "Self-preservation is a form of self-love that ranks at the top of the list of reasons why people lie."[4] And as for greed, that reason for ethical deviation might be best illuminated by the television show *Survivor*.

The CBS show drops a diverse group of people in a remote area, where each competes against others to become the last person standing and win $1 million. They form teams of like-minded (or like-motivated) people. Spoiler alert: This is a phenomenon of the *Deepwater* case covered in Chapter Seven. Inevitably, the competitors must challenge their own moral character to win the $1 million, and they will often discuss their moral dilemmas with the moderator of the show.

What seems to happen every season is that every *Survivor* participant compromises his or her own moral values; the incentive to lie and conspire against others is, of course, a $1 million prize.

But what if the prize were an annual salary of $1.6 million and a $17 million pension, such as the "reward" that former BP CEO Tony Hayward received? That's a little more profitable than surviving poisonous snake attacks by making a $1 million deal with someone who knows how to kill poisonous snakes.

The Macondo blowout was no game. Human lives were at stake. Businesses that supported jobs that supported families were at stake. Long-term environmental harm was at stake.

A total of eight Macondo investigation reports were published—if we look at the two, very different volumes of one joint report as distinct and separate—and all of them in some way rushed to conclusions, with intent or with naïveté, that omitted critical scientific and technical data. None of the six investigation teams creating those reports had legal authority to subpoena witnesses, and six important witnesses to the blowout chose not to testify at hearings—four were supervisors or engineers on *Deepwater Horizon* on April 20. Therefore, all critical facts about the cause of the Macondo blowout were never revealed to accident investigation teams.

Herewith, then, are the five big lies.

Lie 1: People on the rig failed to keep onshore supervisors informed.

In the BP Investigation report, this damning sentence appeared on page 89: "The investigation team has found no evidence that the rig crew or well site leaders consulted anyone outside their team about the pressure abnormality."[5]

That was a deliberate lie by the BP team. Don Vidrine spoke with Mark Hafle, BP Senior Operations Engineer, at 8:52 p.m., fifty-seven minutes before the rig fire. They discussed observed pressures of the second negative test, among other things. Mark Hafle could have decided that remedial action was warranted, but he didn't. We later learned he was on a Continental Airline webpage selecting seat assignments on an airplane during the ten-minute conversation with Don Vidrine.

Mark Bly, BP Accident Investigation Leader and Steve Robinson, investigation team leader, admitted under cross-examination that they were aware of the 8:52 p.m. telephone conversation, but chose to omit this information from the final BP report.

If the BP Bly team had included the telephone discussion in the report, it would have been impossible to assert that well site leaders operated without the full knowledge of onshore supervisors on April 20. It would have undercut BP's ability to place blame on Don and me.

Lie 2: The BP Macondo well team provided operational guidelines for the negative-pressure test.

No, they didn't.

The BP final report states "The BP Macondo Well team provided broad operational guidelines for the negative-pressure test. The rig crew and well site leaders were expected to know how to perform the test."[6]

I was a volunteer five-day substitute to the Macondo team on *Deepwater Horizon*. Of all people, you would think that the new guy on the rig would be handed guidelines, if they existed, so he would be in step with other crewmembers. I was not provided any "broad operational guidelines for negative testing"—and the Bly team knew that. There was no sentence, or even a footnote, stating that five-day well site leader substitute Robert Kaluza was not provided test guidelines.

Digging more deeply into the deception, I discovered that BP had no standard procedures for negative testing, nor did the company provide formal training for negative testing. Facts concerning "training" would become very important later in investigations.

Any curious person covering the story might say, "Show me those guidelines that Kaluza supposedly didn't follow." The Bly team failed to define the "broad operational guidelines," much less provide a copy of them.

The second part of the statement also falls short of the facts: "The rig crew and well site leaders were expected to know how to perform the test." That statement deflected responsibility from BP process safety systems and the Macondo drilling team and directed it toward the well site leaders.

After arriving on the *Deepwater Horizon*, I learned that the rig crews conducted negative tests using a procedure called the "Murry Method" and it had worked well for them. Apparently, this was named after Murry Sepulvado, a *Deepwater Horizon* well site leader who designed it. The Macondo drilling team in Houston had not explained the "Murry Method" to me—I'd never even heard of it until I got to the rig—and that same wells team did not provide a description of a specific negative test plan before negative testing started.

If the Bly team planned to produce an honest investigation report, they would have written this: "BP had no standard procedures for negative testing, provided no negative testing training to employees, and the BP Macondo drilling team did not provide a detailed negative test plan to anyone on the rig prior to conducting negative tests."

Better yet, they could have borrowed the homework done by the National Commission on the BP *Deepwater Horizon* Oil Spill and Offshore Drilling, which stated in its Report to the President released January 2011:

- First, there was no standard procedure for running or interpreting the test in either MMS regulations or written industry protocols. Indeed, the regulations and standards did not require BP to run a negative-pressure test at all.
- Second, BP and Transocean had no internal procedures for running or interpreting negative-pressure tests, and had not formally trained their personnel in how to do so.

- Third, the BP Macondo team did not provide the Well Site Leaders or rig crew with specific procedures for performing the negative-pressure test at Macondo.[7]

Lie 3: The pressure reading meant communication with the reservoir.

The final BP Bly report stated: "There was 1,400 psi on the drill pipe, an indication of communication with the reservoir."[8] That's more than untrue; it's absurd. By irrefutable science, the source of the 1,400-psi pressure observed in the drill pipe was not from a Macondo reservoir.

A version of this statement preceded it, as a kind of warm-up to the notion that the well site leaders on *Deepwater Horizon* were incompetent: "This pressure of 1,400 psi on the drill pipe was misinterpreted by the rig crew and the well site leaders."[9]

It was the BP Bly team who misinterpreted the 1,400 psi on the drill pipe—or fabricated an interpretation that fit their needs. There are five scientific reasons that the source of the 1,400 psi cannot be an indication of communication with the reservoir. The BP Bly team failed to address four of the five scientific reasons.

To their credit, the Bly team did address one of the five scientific reasons why the 1,400 drill-pipe pressure was not from a reservoir pressure and got it right:

In this situation, where the drill pipe should have been in direct pressure communication with the kill line and there was 1,400 psi on the drill pipe, flow should have exited the kill line when the kill line was opened to the mini trip tank. However, the rig crew observed no flow into the mini trip tank.[10]

The above statement is scientifically correct. The kill line did not flow and a rational person would immediately conclude that, because the kill line did not flow, the source of the 1,400 psi drill-pipe pressure could not have been caused by communication with Macondo reservoir.

Instead of this conclusion being the end of a discussion, however, it was just the beginning. The Bly team got creative. They theorized that possibly the kill line was plugged and therefore fluid could not flow through the kill line. Their theory had its roots in desperate imagination rather than science.

The BP Accident Investigation was never intended to be a forum for wild speculation, but in order to find an operational cause for the blowout, the Bly team leaders were forced to make things up.

The Bly team theorized that, after the first displacement and after the annular rubber was closed, 50 barrels of the pit wash/LCM interface fluid leaked past the closed annular rubber and settled in or near the 3 1/16-inch kill line outlet and plugged the kill line. Because the kill line was plugged, flow was blocked and for that reason the kill line did not flow.

The plugged kill line theory is a great example of the embarrassing engineering incompetence exhibited by the Bly group. Here is why:

- We know the annular rubber could not have leaked 50 barrels because the 50 barrels had nowhere to go. Where could 50 barrels of fluid go? It can't go *down* the hole.
- BP hired an independent laboratory to test whether pit wash/LCM interface fluid could indeed plug the kill line. The test results showed that pit wash/LCM interface fluid "could not plug the kill line from transmitting pressure."[11]
- During the preparation for the May 26, 2010 top kill operation, the kill line was cut, and there was no evidence of a blockage in the line at the point where the line was severed or when pumping operations were conducted. The Bly report came out more than three months later.

Without a doubt, the Bly team leaders knew the kill line was not plugged and the 1,400 psi was therefore not caused by communication with reservoir pressure. They had a mandate to find an operational cause for the blowout, however, so they published the debunked theory that the kill line was plugged in the final accident report.

Lie 4: BP failed to state explicitly that the negative test was successful.

This is a lie of omission.

In the final Bly report, the investigation team accurately described the first negative test and provided a graph indicating the timeline for the test and what the test indicated. Unfortunately, anyone looking at the graph who is unfamiliar the kind of testing that would be done on an offshore rig would see nothing more than orange, green, and blue lines going up and down. The report provides some narrative background on the test, but stops short of confirming test success:

> The rig crew began the test by monitoring the drill pipe using the method consistent with their regular practice on prior wells. However, the MMS Application for Permit to Modify for the Macondo well temporary abandonment stipulated that the test should be conducted by monitoring flow from the kill line.

> The well site leader observed the discrepancy and stopped the test. After discussion between the well site leader and the rig crew, the test was resumed by closing the drill pipe and lining up on the kill line (i.e., creating a flow path from the kill line to the cement unit).[12]

The second paragraph doesn't contain a lie, but it does contain a clear error. The flow path of the kill line negative test was aligned to the mini trip tank, not to the cement unit.

Bly team engineers knew well integrity had been established by the first negative test's results but decided not to publish those results. That would make the well site leaders look competent and undermine BP's ability to blame us for catastrophic mistakes.

During the second negative test, the rig crew arranged piping on the rig floor to direct flow through a similar-sized conduit line called a kill line, so flow would go to a small tank called a mini trip tank. The mini trip tank has electronic calibration so the volume that flows into the tank can be observed at a gauge in the drillers shack. As with the drill-pipe test, the kill-line test resulted in a no-flow result. There was no flow for thirty minutes, which meant the casing and the float valve were holding any potential reservoir pressures. Well integrity was established for the second time.

Many experienced and knowledgeable rig supervisors on *Deepwater Horizon* went to bed right after the two successful negative tests. No one goes to bed if there is trouble with an oil and gas well.

Lie 5: Pressure testing performed while the BOP was installed on the Macondo well appeared to conform with BP standards.

Deepwater Horizon's Blowout Prevention (BOP) equipment failed and that failure caused the tragedy. There is no escaping this truth: The BOP equipment was not properly maintained and inspected because senior managers at BP and senior managers at Transocean made deliberate and reckless decisions to delay proper BOP maintenance and inspections. They did not comply with US Code of Federal Regulations 250.446 BOP Maintenance and Inspection Requirements.

A point that has been referenced but not discussed to this point is that the BOPs on *Deepwater Horizon* were not properly pressure tested. Blowout prevention equipment must be regularly pressure tested while drilling an oil and gas well. The principle reason is to ensure that if a kick is encountered the BOPs will seal the pressure of the well kick.

In 2009, the US Code of Federal Regulations governing oil and gas drilling in the Gulf of Mexico required that BOP variable bore rams, blind shear rams, and all high-pressure valves on BOPs be regularly pressure tested to their rated working pressure or maximum anticipated surface pressure, plus 500 psi.

The BOP equipment on *Deepwater Horizon* had a rated working pressure of 15,000 psi. That means the high-grade steel shell of all the closing sections of the BOP stack, flanges, bolts, and all connected valves to the BOPs are rated to contain and hold 15,000 psi of well pressure. The BOP variable bore rams, blind shear ram and casing shear ram were rated to hold 15,000 psi and all choke and kill line valves attached to the BOPs were also rated to hold 15,000 psi. Almost every piece of equipment on the BOPs on the *Deepwater Horizon* was rated to hold 15,000 psi with just a couple of exceptions. Generally, BOP annular rubber elements are rated to contain less pressure; annular rubber elements cannot contain pressures as high as those contained by steel-bodied closing rams. On the *Deepwater Horizon* BOP stack, the upper annular rubber element was rated to contain 10,000 psi and the lower annular rubber element was rated to contain 5,000 psi pressure.

The American Petroleum Institute (API), which crafts standards for the entire oil industry, recommends in RP 53, 18.3.4 that variable bore rams and blind shear rams be regularly pressure tested to "greater than the maximum anticipated surface shut-in pressure." In addition, in its own statements of policies, BP required strict pressure testing standards described in documents such as the *Drilling and Well Operations Practice* (DWOP), specifically section 24.2.2. By strict BP pressure test standards, the BOPs on *Deepwater Horizon* should have been regularly pressure tested while drilling the lowest section of the Macondo well to 9,500 psi.

Key concepts in this discussion of BOP pressure testing are Maximum Anticipated Surface Pressure, (MASP) and

Maximum Anticipated Wellhead Pressure (MAWHP). MASP or MAWHP are worst-case scenarios when an oil and gas well is shut-in after a significant amount of oil and gas has entered a well, or oil and gas has migrated up the well all the way to the BOP stack. The Macondo well was probably producing this exact type of scenario. Oil and gas had either migrated or flowed up the well to just below the BOPs when they were closed, or production casing broke just below the BOPs just before they were closed. There was high oil and gas pressure at the BOPs when the components were activated during the crisis. Drilling managers and engineers understand that a worst-case scenario can occur on almost any oil and gas well, so they require that BOPs be regularly pressure tested to MAWHP.

Drilling engineers and managers must calculate MAWHP before even starting a well because they must properly "size'" the BOP stack to be installed on the well. Engineers get together with geologists and estimate expected formation pressures, then the engineers calculate MAWHP for each hole section. Once geologists and engineers decide the total depth of the well, they look at MAWHP at the deepest depth and select the correctly rated BOP to handle the expected well pressures.

Drilling engineers develop a well plan and advise regulators and drilling personnel involved with the well of the required BOP test pressures for each hole section.

Generally, the deeper the hole section, the higher the BOP test pressures required.

BOPs are generally rated at 5,000 psi increments and must be able to contain all reservoir pressures or there can be a blowout. Additionally, as the well gets deeper, the formation pressures generally go higher, so the BOPs must be properly pressure tested to contain the *highest* pressures deep in the well. In upper sections of the well, it is safe and prudent to regularly pressure test BOPs to 5,000 psi, but in deeper well sections, the BOP equipment must be regularly pressure tested to much higher pressures. The pressures in deepest formations of the Macondo well were approximately 12,000 psi.

The MMS demanded, API recommended, and BP required by strict policy that subsea BOPs be regularly pressure tested to MAWHP. But the BOPs on *Deepwater Horizon* were not.

On April 20, 2010, it would have been anyone's guess if the BOP high pressure equipment could contain and seal a worst-case scenario well kick. The high-pressure BOP equipment such as the variable bore rams and blind shear ram were not regularly pressure tested to MAWHP in the deepest hole section of the Macondo well. They were regularly pressure tested while drilling the deepest hole sections of the well to only 6,500 psi—not even close to MAWHP. The importance of this shoddy testing procedure is something I'll cover a bit more in Chapter

Nine, which features a dramatic testimony on the BOPs during my trial.

Impact of the Lies

Any statements in the BP Final Accident Report are critical in the story of *Deepwater Horizon* because the five other major reports were, to varying degrees, premised on its contents. Challenges to the false assertions, no matter how scientifically baseless they were, were generally weak or non-existent with the notable exception of Volume I of the two-volume Joint Report by the United States Coast Guard and Bureau of Ocean Energy Management, Regulations and Enforcement (BOEMRE). In USCG's Volume I, the Coast Guard investigative team had an obvious intent to get the story right.

Human beings tend to absorb as truth something heard or read over and over again no matter how baseless it is. We think cavemen lived in caves, watching TV close to the screen will damage our eyes, people have only five senses, and bats are blind. When someone challenges those well-established "truths," that person is judged belligerent rather than fact-based. Because people with the "right" credentials put their names on the BP report, the falsehoods therein morphed into supposedly reliable information that kept getting repeated—and believed.

The BP Accident Investigation Report reflected a mandate from the top to determine an operational cause of the blowout, so it was not meant to be a document with scientific or engineering integrity. It was written to cover-up BP management systems failures. I would call it one of the greatest cover-your-ass documents ever produced by a company.

Seeing Through the Lies—Sometimes

The Transocean Accident Investigation Report was published June 2011, nine months after BP's. The false and misleading statements published by the Transocean investigation team built on some of BP's, but they were not designed to indict fellow employees.

Driven by an agenda to find the truth rather than protect anyone, the US Coast Guard poured through the evidence and arrived at conclusions that undercut some key assertions by BP and Transocean. In other words, USCG tried to find out if bats really are blind.

The USCG published their Macondo Investigation Report on September 9, 2011, and they should be commended on many levels. The USCG team carried out an excellent effort of investigating the working conditions and maintenance of many types of equipment onboard the *Deepwater Horizon*. Most notably, the USCG investigation team was the only team to identify the single direct cause of the Macondo blowout. Volume I of their report, which addressed areas of USCG responsibility and was therefore focused mainly on Transocean rather than BP, stated:

One of the more serious maintenance issues identified during this April 2010 audit related to Transocean's BOP, manufactured by Cameron. The report stated that "upon review of certification documentation it was noted that the date of last manufacturer's certification was 13 December 2000" and "this is beyond the 5 yearly [sic] inspection, overhaul, and re-certification requirement." Rather than follow the American Petroleum Institute Recommended Procedure (API RP) 53, which called for inspection and certification every three to five years, Transocean had decided to use what it called a "condition-based" maintenance program, which did not require inspections on any particular schedule.[13]

Condition-based maintenance is a reckless, misguided maintenance system that second-guesses the manufacturer's recommendations, API recommendations, and the US Code of Federal Regulations. By admitting to investigators that they used a condition-based maintenance approach, Transocean managers chose a lesser-of-two-evils defense strategy. It seemed better than admitting that the BOP equipment on *Deepwater Horizon* was not properly maintained for years.

The USCG Team got even more pointed and colorful in its criticism of Transocean in this summary statement:

Transocean's safety management system had significant deficiencies that rendered it ineffective in preventing this casualty. The company leaders' failure to commit to compliance with the International Safety Management Code and created a safety culture throughout its fleet that could be described as: "running it until it breaks," "only if it's convenient," and "going through the motions." This is best illustrated by the condition based maintenance of the BOP, and the deferral of recertification and required maintenance, the bypassing of alarms and emergency shutdown devices, and the conduct of emergency drills. This culture resulted in poor materiel conditions, ineffective decision making, and inadequate emergency preparedness for responding to catastrophic events.[14]

Transocean's own report quickly dismissed the possibility they had failed to maintain the BOP: "Forensic evidence from independent post-incident testing by Det Norske Veritas (DNV) and evaluation by the Transocean investigation team confirm that the Deepwater Horizon BOP was properly maintained and operated."[15] That's like saying, "Pigs can fly."

The Bureau of Ocean Energy Management, Regulation and Enforcement (BOEMRE) produced Volume II of the joint report with the USCG, but it has some serious flaws that were not characteristic of the USCG report. BOEMRE was formed three months after the Macondo blowout when the US Department of the Interior split the Mineral Management Service (MMS) into three parts to avoid the conflicts of interest that had plagued MMS. BOEMRE criticized BP in its report; however, it reflected some major mistakes. One problem was that BOEMRE, like Transocean, used DNV to do a forensic examination of the BOP stack. I explore this fox-in-the-henhouse scenario a bit more in Chapter Seven.

The BOERME investigation team also did not reveal in its Investigation Report that MMS managers and rig auditors failed to enforce 30 CFR 250.446 BOP maintenance and inspection requirements from 2006 through 2010, which was a direct cause for the Macondo blowout.

The most egregious false statement in the report, however, was that the BOERME investigation team implicated BP well site leaders and Transocean rig members for "failing to recognize risk."[16] That comment led to an enumeration of problems that supposedly escaped the notice of the well site leaders, OIM, and toolpushers—even though we had about 100 years of combined experience doing jobs on offshore rigs. BOERME's observations were dangerous in terms of public perception because the well-intentioned team was not populated with experienced deepwater drillers or engineers. The crude way of saying it is that they just didn't know what they were talking about.

Unfortunately, it was statements like this that supported the Justice Department's indictments of Don Vidrine and me on November 14, 2012.

The goal of the BOERME investigation team was to examine the relevant facts and circumstances concerning the root causes of the *Deepwater Horizon* oil disaster. I am certain no one instructed the BOERME investigation team to make subjective comments, yet this comment crept into the report:

> Testimony and interview notes from BP personnel revealed that they had an oversimplified view of what constituted a successful negative test – they each believed that they only had to check for flow to evaluate whether a negative test had been successful.[17]

With more than a half-century of experience between us, it's unlikely that Don Vidrine and I had an oversimplified view of what constitutes a successful test. The BOERME investigative team did not understand the factors involved in the test, and that rig supervisors and rig crews follow approved well plans unless there is strong disagreement with the plan. The Macondo well plan stated that we should monitor the well for 30 minutes to ensure no flow. The only criterion for success or failure of a negative test is flow or no flow. That is not an oversimplified view; every negative test ever conducted on *Deepwater Horizon* required that same criterion of success or failure—flow or no flow.

The BOERME investigation team not only made numerous false statements out of ignorance, but they also made alarming subjective judgments like the one above and published them in their final investigation report.

Sometimes, the investigation reports contained information that made me scratch my head and wonder, "Where the heck did they get that?" The errors were not even cut-and-paste mistakes, but complete misunderstanding of whatever evidence the teams were reviewing. At times, though the reason for is that investigators were personally given incorrect information.

A case in point is from the National Academy of Engineering and National Research Council report of December 14, 2011: "The negative pressure test was attempted three times, as described in the BP accident investigation report (BP 2010)."[18]

This was an investigation team that invited the BP Macondo wells team into their hen house and it showed in their report. The BP Accident Investigation Report described two negative tests, not three, but BP executives had been invited to make a presentation to the Academy and this "information" about three negative tests probably sprouted from it.

I am not surprised that the National Academy team was hoodwinked. Other teams that allowed foxes in the hen house were also hoodwinked.

And now, presenting the most absurd attempt by a scientific body at presenting analysis of the *Deepwater Horizon* blowout. This is from the Chemical Safety Board in their report of June 5, 2014.[19] In the Overview, this statement appears in a footnote:

The BOP is not the most critical barrier to loss of well control, yet, as will be explored in Volume 2, industry had placed great reliance on the effectiveness of BOPs for preventing major well accidents. The BOP was also a key piece of physical evidence available to the entities that investigated the accident. Much of the physical evidence was not retrieved or was lost with the sinking of the rig, while other physical evidence, including the downhole cement, remains thousands of feet under the seabed.[20]

The BOP stack is the single most important critical barrier to loss of well control. The BOPs must function properly 100 percent of the time or a well kick will turn into a well blowout. The CSB sounds uninformed—and I am being polite.

There can be many contributing causes of well kicks, such as a bad cement job, a reduction of overbalance pressure due to human error, unexpected high reservoir pressures, and many other contributing causes. There is usually only *one* direct cause of a blowout: the BOP equipment fails. The BOPs on *Deepwater Horizon* failed five times on April 20, 2010.

I want to end this critical look at the major investigative reports with a quick reference to the President's Report and then conclude with the one generated by the same group that most closely drew from the BP document.

The final version of the President's Report written by the National Commission on the BP Deepwater Horizon Oil Spill and Offshore Drilling is a narrative-based account of the disaster with a focus on, as the title indicates "Deep Water: The Gulf Oil Disaster and the Future of Offshore Drilling." As noted above, it does contain some useful conclusions about the conditions and circumstances onboard *Deepwater Horizon*, but it's also very focused on lessons learned that would help shape government policies about offshore drilling.

The Chief Counsel's Report, also generated by the bipartisan Commission established by Barack Obama a month after the disaster, came out in preliminary form in October 2010 and was issued in final form in February 2011. This was an investigative report about the causes of the tragedy.

BP clearly tried to make it easy for the Commission to conduct its investigation and draft its report. The Acknowledgements in the General Counsel's report are almost effusive in thanking BP, noting that BP "provided us access to supporting documents" for their final report and saying that the report itself "aided our efforts, and we commend BP for undertaking it."[21]

They also indicated that members of BP's Macondo well team had explained the chain of events that led to the blowout to them—something the team wouldn't have done with much detail because that would have shed light on the actual cause of the blowout.

The Chief Counsel team did reveal that the *Deepwater Horizon* BOPs were out of CFR 250.446 regulatory compliance for several years and called that fact "deeply troubling"[22] although not cited as causative in the disaster.

A preliminary version of the report was circulated in October 2010 and on November 8, 2010, Chief Counsel Fred Bartlit, Jr. made a presentation to a Presidential Panel in Washington, DC addressing findings from the Chief Counsel's investigation. Bartlit had a rousing opening that smelled of BP propaganda: "To date we have not seen a single instance where a human being made a conscious decision to favor dollars over safety."[23] And in case someone didn't hear him the first time—or didn't believe him—he repeated himself.

If Bartlit knew of any of the dollars-over-safety decisions made by BP in Table 5.2 of the Chief Counsel's report, then he lied to the Presidential Panel. Following is a recreation of the table.

Examples of decisions that increased risk at Macondo while potentially saving time.[24]

Decision	Was There a Less Risky Alternative Available?	Less Time Than Alternative?	Decision Maker
Not Waiting for More Centralizers of Preferred Design	Yes	Saved time	BP onshore
Not Waiting for Foam Stability Test Results and/or Redesigning Slurry	Yes	Saved time	Halliburton (and perhaps BP) onshore
Not Running Cement Evaluation Log	Yes	Saved time	BP onshore
Using Spacer Made from Combined Lost Circulation Materials to Avoid Disposal Issues	Yes	Saved time	BP onshore
Displacing Mud from Riser before Setting Surface Cement Plug	Yes	Unclear	BP onshore

Setting Surface Cement Plug 3,000 feet below Mudline in Seawater	Yes	Unclear	BP onshore
Not Installing Additional Physical Barriers during Temporary Abandonment Procedures	Yes	Saved time	BP onshore
Not Performing Further Well Integrity Diagnostics in Light of Troubling and Unexplained Negative Pressure Test Results	Yes	Saved time	BP (and perhaps Transocean) on rig
Bypassing Pits and Conducting Other Simultaneous Operations during Displacement	Yes	Saved time	Transocean (and perhaps BP) on rig

Having read all the reports, I would say that none of the government-funded Macondo investigation reports contained lies those investigators created—only ones they lifted from other sources. Government-funded investigation teams published false statements and inaccuracies, but none did it with the intent to deceive in my opinion. The government teams were not qualified, trained or experienced deepwater drillers or engineers, therefore the teams were incapable of lying like engineers and managers from BP and Transocean.

In my opinion, the leaders of every government investigation team, the leaders that made final editing decisions of what was published in final reports, were not lying with knowledge, intent, or malice—except possibly in one case, and that was Chief Counsel Fred Bartlit, Jr. asserting there was no evidence that BP put dollars over safety.

Multiple trials, Congressional hearings, and class action lawsuits kept BP and Transocean personnel and their hired expert witnesses testifying soon after the blowout. The problem was that some of the experts—knowingly and unknowingly—presented false information during their testimonies and in their official reports to judges and government authorities. Others compounded misperceptions about the blowout by testifying without knowing much, or anything, about offshore drilling.

The key thing to note is that every Macondo investigation report contained numerous false statements, several theories without scientific basis, and numerous inaccuracies.

- Critically important witnesses—people who were on *Deepwater Horizon*—were not asked for contributions or testimony.
- Some companies failed or refused to turn over requested documents.
- Some investigative teams were not populated with deepwater drilling experts or engineers.
- Many false and misleading statements from the BP Accident Investigation Report were essentially cut and pasted into other investigation Reports.

CHAPTER SEVEN

2 SCAPEGOATS

Scapegoats—that's what BP desperately needed to divert media attention and legal heat from senior corporate officers. BP CEO Tony Hayward led the way in painting himself and the company as victims, strongly indicating he and his senior leadership team did not have the broad shoulders to take responsibility for the disaster. Statements like these by Hayward suggest that BP was run by the kind of people who liked to minimize problems and blame everyone but themselves:

> April 29, 2010: "What the hell did we do to deserve this?"[1]
> May 13, 2010: "The Gulf of Mexico is a very big ocean. The amount of volume of oil and dispersant we are putting into it is tiny in relation to the total water volume."[2]
> May 18, 2010: "I think the environmental impact of this disaster is likely to be very, very modest."[3]

May 31, 2010: "There's no one who wants this over more than I do. I would like my life back."[4] (He got his life back when he was abruptly replaced by Bob Dudley, who took over as CEO the following July and headed the plea deal negotiations that led to the scapegoating exercise.)

These gaffes are nothing compared to the lies told by some of the BP expert witnesses—lies designed to put Don Vidrine and me in prison so everyone at corporate could rest easy that "justice" had been done and they were off the hook.

This chapter focuses on the situation Don and I faced, although I also want to point to the reasons a former BP Vice President, Dave Rainey, had a bull's eye on his back, too.

The important background is that Macondo wasn't BP's first disaster scene and the Department of Justice wanted indictments against flesh and blood BP employees this time around. BP's history of disasters included four major events in the five years prior to the *Deepwater Horizon* blowout.

- On March 23, 2005, an explosion at BP's Texas City, Texas, refinery killed fifteen workers and injured 180 others. At the time, it was one of the worst industrial accidents in United States history. Hydrocarbon liquid and vapor were accidentally released and ignited, with BP admitting that safety procedures at the refinery had been ignored. BP paid over $2 billion in claims settlements, and an additional $71.6 million in fines for worker safety violations and $100 million in fines for the pollution caused by the explosion.[5]

- In March and August of 2006, there were leaks at BP's Prudhoe Bay oilfield in Alaska. The result was the largest oil spill ever in Prudhoe Bay, which is the biggest oil field in the United States. Government officials cited BP for failing to heed warning signs of imminent internal corrosion. In 2007, the company agreed to pay a $12 million criminal fine, $4 million in community service payments, and another $4 million in criminal restitution to the State of Alaska.[6]

- On November 29, 2009, a pipeline at the Lisburne field in Alaska leaked oily material onto the snowy tundra—an event at Lisburne that was repeated less than two years later. Lisburne is adjacent to the Prudhoe Bay field.[7]

Partially due to these previous incidents and overwhelming political pressure from President Barack Obama, the US Congress, and the national news media, the Department of Justice wanted to see bodies in prison—BP bodies. In 2012, the BP Board of Directors obliged and offered to sacrifice three employees to meet DOJ demands. DOJ attorneys accepted the deal, and on November 15, 2012, they indicted two of us who were well site supervisors—not part of the BP senior management planning and engineering team—and a former BP vice president.

Because I was one of the people indicted, I know exactly how the scheme was accomplished. I and my team of attorneys were shocked at the legal action, but in hindsight, it's easy to explain how BP and the DOJ colluded to identify and sacrifice Dave Rainey, Don Vidrine, and me.

Fingering Dave Rainey

Former BP VP Dave Rainey was targeted to appease Members of the US Congress who sat on the House Committee on Energy and Commerce. Those House committee members accused Dave of lying under oath to their committee by under-estimating the amount of oil that flowed from the out-of-control Macondo well after the blowout. Dave used the best data and formulations available to him to give his estimate, but it wasn't what the House committee wanted to hear, so influential members of the House Committee pressured the DOJ to indict Dave Rainey and threaten dire consequences toward BP if they didn't give up Rainey.

As a consequence for Dave's so called lies to the committee, and because the oil spill created such a high degree of harm to the Gulf Coast environment and people, several House committee members threatened to lock BP out of future oil and gas lease sales on all federal lands.

That would have included locking BP out of future oil and gas lease sales in the US Gulf of Mexico—BP's cash cow. A lease sale lock-out in the Gulf of Mexico would be devastating to the very existence and viability of BP as a major oil and gas production company. BP Board members and executives knew precisely the dire consequences they faced if such Congressional action were taken. It would be more than the sharp slap on the wrist they had received after previous accidents; it would be a crippling kick in the groin.

Members of BP's Board of Directors agreed to do almost anything to continue bidding for future oil leases in the Gulf. The possible consequence of innocent BP employees spending years in prison was worth it to them to sustain oil and gas production in the Gulf of Mexico.

I have often wondered if any BP Board member struggled with his or her own morality in agreeing to trade human beings for assurances of continued profitability.

Because Rainey had held a VP role at BP, he was an even better scapegoat target to the DOJ than low level drilling rig well site leaders. The DOJ often announces that it always goes after corporate employees at the highest levels of management. That of course, is an egregious lie explored with intelligence and great documentation by Pulitzer Prize winning author Jesse Eisinger in *The Chickenshit Club*.

Dave's surreal American injustice began when he was indicted by the DOJ on November 15, 2012. It ended on

June 5, 2015 after a twelve-person trial jury deliberated for less than two hours before finding Dave not guilty of all crimes. He walked out of the courthouse without saying a word. Dave is a British citizen and probably went home to England in disbelief over his treatment by the US Department of "Justice."

Regrettably, Dave Rainey, his family and friends were forced to endure two-and-a-half years of stress, anxiety and trauma due to false charges filed by the DOJ simply for political appeasement and with the assistance of his former employer. Furthermore, when the injustice ended in 2015, neither BP nor the DOJ made a public apology to Dave and his family. When the DOJ fails to publicly apologize after they have prosecuted an innocent person, the callous behavior sustains the injustice.

Targeting the Top Two: A Chronology

There are fourteen steps involved in identifying Don Vidrine and me as the people on the rig who were easiest to indict on manslaughter charges and violation of the Clean Water Act.

Step 1: Limit the scope of the BP Accident Investigation.

Within the first couple of days after the Macondo blowout, the BP executive management staff met in Sunbury and Houston to discuss what happened. Damaging facts must have had center stage: Executives were informed that the Macondo drilling team, the drilling team that planned and engineered the Macondo well, broke US federal regulations for four years. The Executives were likely informed that the Macondo team deviated from *Drilling and Well Operations Practices and Engineering Technical Practices* (DWOP) multiple times while planning and drilling the Macondo well. The Macondo well was over budget by more than $40 million and the Macondo drilling team made many decisions that placed dollars over safety.

They were likely informed that BP set the production casing at an awkward depth across open pressured formations and that the Macondo team approved and pumped a very risky production casing foam cement job. In fact, the executives may have been informed that the casing cement job was pumped before all laboratory tests on the stability of the cement were completed by Halliburton, the cementing company.

If the BP executive team was informed of these facts, they were probably shocked by the negligence of the drilling team. A BP drilling team had been in violation of federal law for four years, the Macondo team had deviated from the company's own *minimum* drilling standards multiple times, and the Macondo team disregarded safety if it meant saving time and money. Some executives may have concluded that some BP officials could go to jail for a long time unless they acted quickly.

Whatever initial information BP executives heard, I am certain they promptly discussed a plan to save BP as a company and keep BP managers out of prison.

Here is something that involves no conjecture: We are certain that two or three days after the Macondo blowout and loss of life, BP CEO, Tony Hayward met with Mark Bly, Group Head of Safety and Operations, and appointed him to lead a BP Accident Investigation Team. Mark was directed to severely limit the scope of the investigation by avoiding any probe into BP management's role in the disaster and, instead, concentrating the entire investigation on "immediate or operational" causes of the accident.

The constraints on the so-called investigation were not a secret; this was a fact revealed by BP's safety chief and reported by multiple news sources covering the aftermath of the blowout. From the *Greater Baton Rouge Business Report,* February 28, 2013:

> BP deviated from its own policy on responding to accidents by not determining how management contributed to the 2010 Gulf of Mexico oil spill, a company safety expert who led an internal investigation of the disaster testified this morning. Mark Bly, who was promoted to an executive management position at BP after the report was issued, said he and former BP CEO Tony Hayward decided on the scope of the investigation days after the Macondo well blowout and Deepwater

Horizon rig explosion off Louisiana's coast. Plaintiffs' attorneys at a civil trial in federal court in New Orleans are trying to show BP's report was self-serving, incomplete, and designed to shield the company from billions of dollars in damages that are a subject of the trial. Bly told U.S. District Judge Carl Barbier that he could have gone further in the probe but that there were limitations in terms of availability of witnesses and information. Bly said that, among other things, the probe didn't analyze the potential that cost-cutting contributed to the blast. He said it was decided to limit the probe to operational causes rather than include systemic causes, even though company policy at the time required both factors to be considered.[8]

From *The Times-Picayune*, also on February 28, 2013:
The internal report also did not probe whether BP executives were to blame for mistakes leading to the accident. Bly also acknowledged that the report chose not to explore potential systemic problems with the oil giant's safety policies or failures within its management system.[9]

By narrowing the scope of the accident investigation, CEO Hayward was deviating from BP policy that required a broad, comprehensive scope of investigation.

Step 2: Assemble an investigation team that would follow orders.

BP executives assembled an investigation team willing to narrow the scope of the accident investigation and exclude 100 percent of the BP managers and engineers who planned, engineered and sent approved well plans to the *Deepwater Horizon* from investigation. As a corollary, they were willing to concoct arguments that pinned the blame on fellow BP employees.

The team predominately consisted of long term and loyal BP managers and engineers. BP attorneys advised them about how to handle documents, notes, and witnesses. They schooled them on how to write the final BP Accident Investigation Report.

In the report itself, the *Terms of Reference*, which defined the scope of work, indicated that the investigation should "identify critical factors (events or conditions that, if eliminated, could have either prevented the accident or reduced its severity) and examine potential causal or contributory factors at the immediate cause and system cause levels."[10] That couldn't happen if the team adhered to their mandate to narrow the scope of the accident investigation.

Mark Bly followed CEO Tony Hayward's direction and so did his team. Their investigation covered only immediate or operational causes and excluded, completely, investigation of BP management.

The approach enabled Hayward to cover-up systemic BP management failures—deliberate and reckless failures that led directly to the blowout. Hayward had effectively handed out free passes to BP managers and senior engineers.

The flip side of Hayward's directive was that the entire BP accident investigation focus, effort and concentration would be directed toward two BP employees onboard the rig—employees that had done no well planning, no engineering, and made none of the decisions to violate any Code of Federal Regulations or federal law. Members of the team dutifully kept their focus on what happened on the rig and turned their backs on whatever had transpired in BP's offices and conference rooms. The conclusion of the so-called investigation was pre-determined before the team ever got started: They would cite operational causes for the blowout.

Step 3: Shut out the best witnesses.

Following the Hayward directive, the Bly team focused its entire investigation on "immediate or operational causes" that may have led to the Macondo blowout. The best available witnesses for the Bly team would have been crewmembers who were on *Deepwater Horizon* in the days, hours and minutes before the blowout. Prime witnesses for the Bly team would have been the rig site BP well site leaders—Don Vidrine and me.

If consulted, people on the rig would have been able to tell the Bly team where specific people were on the rig at specific times, the timeline of events would have been precise, and the Bly team could have discovered how decisions were made and who made them. They would have also learned whether or not anyone on the rig had safety or operational concerns, and if anyone had wanted to "stop the job" due to safety concerns. Continuing to dive deep, they would have discovered what well plans were followed on the rig, who communicated with engineers in Houston and at what time, whether or not tests conducted on the rig were successful, what assessments were of the recent cement job, the integrity of the dual floats, who was in bed, what drilling rig equipment was inoperable at the time, and details of many more important issues. Most importantly, the Bly investigation team could have asked both BP well site leaders to read the final draft of their final investigation report before publication to ensure the final BP accident investigation report was accurate. Don and I were kept beyond arm's length, however, so we could not do anything to help the team obtain accurate data and information.

During the entire five-month investigation, the Bly team only once contacted me for information or to confirm any type of accurate assessment of what happened. I met only once with four members of the Bly team for about three hours within two weeks after the blowout. They never contacted me again.

For someone who ultimately had such a central role in the *Deepwater Horizon* legal proceedings, you would think that what I observed and experienced would have been of more interest.

In the opening notes, the BP Accident Investigation Report states, "This report is based on the information available to the investigation team during the investigation."[11] Since the Bly team made a conscious choice not to include certain sources, or not to revisit initial contributions from key witnesses, that statement is not true.

Step 4: Submerge damning evidence.

There was a strategic reason why the Bly investigation team chose to avoid follow-up contact with key witnesses such as Don Vidrine and me to verify facts. We were among the team's best sources of information about the actions and decisions immediately prior to the blowout that were under scrutiny. BP executives and Bly team leaders aimed to tell a BP story, not the entire true story about the events that led up to the blowout. The BP story, of course, was a deliberate cover-up story.

As a quick aside, let's look at how accustomed many of us have become to corporate spin, thanks to the effective efforts of public relations professionals. The relevance is that the BP Accident Investigation Report is essentially a PR document.

My co-author, Maryann, recalls a time when she was doing public relations for a high-technology industry trade association and some major companies had already initiated layoffs, with more layoffs looming in the near future. One of her superiors instructed her to issue a press release about the industry's increased productivity. The basic message was, "We're still making a lot of money, but doing it with fewer people." There was to be no mention of layoffs. In a world where spin supports profits and profits lure happy shareholders, companies invariably minimize their mistakes and problems, or don't mention them at all.

Bly investigation team leaders did not ask the BP well site leaders to read the final draft of the accident investigation report before publication to ensure accuracy and honesty because the team had intentionally hidden the key facts about the cause of the blowout. There is, of course, a bigger "because;" that is, they intended to craft a report that made the BP well site leaders complicit in the catastrophe.

The final BP Accident Investigation report contained many false statements and inaccuracies about the events and decisions that led up to the blowout—the subject of coverage in other portions of this book. It also avoided suggesting the actual cause of the blowout, which is a lie by omission. Even hinting at the non-compliance of the BOPs would have implied recklessness on the part of BP senior management, and that would have been a violation of their mandate to keep the report narrowly focused on operational causes.

Step 5: Have well site leaders' testimony omitted from hearings.

When witnesses who are generally perceived to be central to a case do not testify at hearings or at a trial, a common assumption might be that the witness has something to hide. People may assume when they read that a witness isn't testifying that some lawyer made a deal to keep this person—possibly a guilty party—off the witness stand. In the Macondo proceedings, that couldn't have been further from the truth.

Attorneys must weigh a legal strategy versus public opinion when they advise a client not to testify.

On May 10, 2010, BP hired an attorney team to represent me during the long Macondo legal journey. I was assigned Shaun Clarke, who added his partner David Gerger and their colleagues Dane Ball and David Isaak to the team. Their offices were in Houston, Texas. Since I live just outside of Las Vegas, most of our communications were done over the phone. Despite the geographic separation, I felt they were right beside me, at least in spirit, through the entire long journey.

All four attorneys showed impressive interest, diligence, and intelligence. But they weren't scientists or engineers, so when I began explaining what we did onboard *Deepwater Horizon*, I spoke a foreign language.

Just two weeks after I was assigned attorneys I was scheduled to testify at the Marine Board of Investigation, MBI, hearing in Kenner, Louisiana on May 27, 2010 concerning the Macondo blowout. Before that date, I was able to meet only twice with Shaun Clarke and David Gerger in Houston where we had introductory conversations concerning legal issues and what happened on the *Deepwater Horizon*. Offshore deepwater drilling is a very complex activity that cannot be taught and understood overnight so when I explained what happened during the four-and-a-half days while onboard the *Deepwater Horizon*, I think my attorney team was somewhat overwhelmed by the amount of technical information. Deepwater drilling cannot be taught and understood well in a few sessions or a few months—not even in a few years. Deepwater drilling is very complex; it is helpful to have a science background to understand it.

Realizing the complexity of the issues and the distortions about the blowout that were taking shape, they advised me not to testify at the MBI hearing.

The decision not to testify involved two consequences. First, some media coverage surrounding the hearing had churned convictions that I was probably guilty of something. Second, my attorneys realized that if I got caught-up in the political vortex of the Macondo accident, I might face legal issues later, and that whatever I said at a hearing could be twisted and used against me. Additionally, not yet grasping the complexities of deepwater drilling and what we faced on *Deepwater Horizon*, they were not able to understand the complete story of what happened, therefore, they may have felt they could not effectively advise me at the MBI hearing.

At least one attorney on the team understood in 2010 that the US Department of Justice under Eric Holder was highly politicized. Knowing that, it was easy to surmise that someone from BP would likely be indicted. Despite my desire to testify, he felt that not testifying at the MBI hearing would keep me out of the spotlight. At least for the moment.

I had nothing to hide. I was anxious and willing to testify, but the decision I made on attorney advice to forego testifying at hearings created a bias within Macondo investigation teams that may have made it easier to target me and Don Vidrine for later recrimination.

Step 6: Continue to control the information (that is, get the foxes in the hen house).

Published September 8, 2010, the BP Accident Investigation Report was the first investigation report released after the blowout. The second was The Report to the President, published January 11, 2011.

By publishing first, BP received myriad benefits. Other investigative teams read the BP report and used it as a reference source in discussing certain events and theories without sufficient fact checking or scientific challenges.

The Macondo team members were like foxes sneaking into many hen houses.

At no presentation or hearing did the Macondo well team reveal that the BOP equipment onboard *Deepwater Horizon* was out of Code of Federal Regulation 250.446 BOP Maintenance and Inspection compliance for four years—a fact that meant BP had been in violation of the law since April 2006. The team also failed to inform other investigation teams that the Macondo drilling team that planned and engineered the Macondo well deviated several times from BP Policy (DWOP) and MMS Policy, thereby setting the stage for the blowout. If the Macondo wells team had presented a complete, truthful story there would not have been any scapegoats.

The National Commission on the BP Deepwater Horizon Oil Spill and Offshore Drilling, which had produced the Report to the President, also created the Chief Counsel Report, released February 2011. They seemed enamored with the Macondo wells team's story and even commended BP personnel for their efforts.

Many individuals involved in the Macondo incident spoke with us voluntarily. They included rig crew members, cementers, mudloggers, equipment suppliers, and shore-based engineers. Notably, many members of BP's Macondo well team met repeatedly with us to explain the chain of events that led to the blowout. BP also released a report of its own nonprivileged investigation of the blowout, and then provided us access to supporting documents. While BP's report differs from ours in scope, purpose, and conclusions, it aided our efforts, and we commend BP for undertaking it.[12]

Chief Counsel Fred Bartlit went well beyond this polite commendation in defending BP when he said in relation to his Macondo investigation: "We have not seen a single instance where a human being made a conscious decision to favor dollars over safety."[13]

In late 2010, after the BP report came out, the Bureau of Ocean Energy Management, Regulation and Enforcement (BOEMRE) hired Det Norske Veritas to lead the Deepwater Horizon BOP investigation after the BOP stack was pulled out of the ocean and transported to a dock in New Orleans, Louisiana. DNV is the rig audit company that gave *Deepwater Horizon* a pass for several years after the BOP went out of compliance.

DNV chose to assemble a five-member Technical Working Group (TWG) to assist with the BOP inspections that met each day to analyze and plan future tasks of investigation of the BOP. DNV invited employees from Transocean, Cameron, and BP into the TWG. Not to belabor the foxes in the hen house analogy, but this is such a blatant example, it's easy to envision the foxes drooling at being in the BOP hen house.

The Macondo wells team was later invited to make a presentation of the "BP Story" to the National Academy of Engineering and National Research Council Investigation team and to the DOJ. The foxes were eating well.

Two members of the Macondo wells team, BP investigation team leader Mark Bly and BP VP Steve Robinson, were witnesses that testified for BP at the oil spill civil trial in New Orleans presided over by Federal Judge Carl Barbier. In the Phase One Trial "Findings of Fact and Conclusions of Law," Judge Barbier ruled that both Bly and Robinson were responsible for providing "patently false"[14] information in his court. Barbier concluded, "The explanations provided by Messrs. Bly and Robinson are untenable."[15] The Judge also did not curb his opinion of the editorial liberties taken in the Bly report when he stated:

The Court infers that BP's investigation team recognized the importance of the 8:52 p.m. phone call and chose to omit it from the BP Accident Investigation Report to avoid casting further blame on BP.[16]

BP managers Mark Bly and Steve Robinson were caught in a lie in a courtroom under oath. Months earlier, they had ample opportunity to practice their lies in the presentations to other Macondo investigations teams and at the US Department of Justice.

In contrast, the Macondo investigation teams and the DOJ that invited the BP predators into their house did not challenge their story. They accepted the cover-up that would hold up as long as BP could convince anyone who would listen that two well site leaders were the villains.

Step 7: Find an operational cause.

We now know the single direct cause of the blowout was failure of the BOPs to seal the kick. That fact was not made clear until my trial in February 2016, however.

As directed, the Bly accident investigation team focused its entire investigation on immediate or operational causes. The team therefore looked carefully to find any operational cause that could have contributed to a blowout. During investigation the Bly team discovered an odd pressure on a drill pipe and homed in their full attention on that unexplained pressure. Maybe that unexplained pressure could be the operational cause, the smoking gun, they needed to blame the disaster on someone onboard. And that is why Bly and his team directed extraordinary attention to drill pipe pressure related to one of the negative tests as a possible cause of the blowout.

BP Bly team investigators stated that the 1,400-psi drill pipe pressure I described in Chapter Three was "an indication of communication with the reservoir"[17] in its final Accident Investigation Report. The team provided no scientific proof of their assertion but failing to investigate any other possible explanations for the drill pipe pressure, they concocted an explanation out of their hypothesis.

The Bly investigation team theory that the source of 1,400-psi drill pipe pressure was created by Macondo reservoir pressure is scientifically baseless. In fact, there are five scientific explanations why the drill pipe pressure cannot be the result of communication with the reservoir. Yet, the Bly team had found their much sought after operational cause and therefore ignored facts that proved them wrong.

Of the five scientific explanations why the 1,400-psi drill pipe pressure could not be communication with the Macondo reservoir, the simplest to understand is the one that follows.

At 8:00 p.m. on April 20, 2010, the 1,400-psi pressure that registered during the second negative test was bled off the drill pipe. The pressure bled off immediately. In less than sixty seconds, pressure on the drill pipe went to zero pressure and zero flow. If the 1,400 psi had been created by communication with a Macondo formation pressure, as theorized by the Bly investigation team, when the drill pipe was opened at 8:00 p.m., flow out of the pipe would have been immediate, massive and continuous; flow would have never stopped! In fact, flow out of the drill pipe would have looked very similar to the oil and gas flow that continue for eight-seven days from the out-of-control Macondo well. The Bly investigation team ignored the scientific proof.

And then there's common sense, mentioned before as a relevant factor in countering the "operational cause" promoted by BP. No drilling supervisor would ever go to bed if there was any problem with an oil and gas well. The BP Bly team never considered common sense because it would have challenged their assertions.

Step 8: Upsurge of interest in indictments.

Loss of human life had also occurred at BP's Texas refinery in 2005, but the massive damage to the Gulf of Mexico caused by the Macondo blowout evoked a frenzied response from environmentalists. In the minds of many people, someone had to pay for the environmental catastrophe and money alone was insufficient payment. In late 2011, one year before the US Department of Justice indicted Don Vidrine and me, University of Michigan Law Professor and former DOJ attorney, David Uhlmann, wrote: "It is unlikely BP, Transocean, and Halliburton would face criminal prosecution but for the tragic harms caused by the Gulf oil spill."[18] Uhlmann served for seventeen years at DOJ with the last seven as chief of the Environmental Crimes Section.

Uhlmann thought through the process of targeting Don and me without mentioning us by name. In his June 2011 paper, "After the Spill Is Gone: The Gulf of Mexico, Environmental Crime, and the Criminal Law," he goes through a process of eliminating likely candidates for prosecution, and then arrives at the remaining prospects—rig supervisors. The problem is that he made a giant assumption about how unlikely it was that far-away senior executives could be culpable:

> Absent false statements or obstruction of justice, however, the Justice
> Department may struggle to identify culpable individuals who possessed sufficient management authority in the Gulf oil spill. Unless the government departs from its prior practice and charges strict liability violations of the Migratory Bird Treaty Act, only those directly involved in the oil spill can be charged with crimes. To charge individuals with a criminal violation of the Clean Water Act-the primary statute for addressing the spill-the government would need to show that the defendants acted knowingly (for felony charges) or negligently (for misdemeanor charges). Yet it is unlikely that senior executives of BP, Transocean, and Halliburton, who had the greatest influence over the corporate culture that made the spill possible, played such a personal role in the disaster.

The question therefore becomes whether the government can identify individuals with enough supervisory responsibility and personal involvement to be blamed for the Gulf tragedy.

The president's commission on the Gulf oil spill identified a number of shore-based engineers, supervisory personnel on the rig, and rig workers who were involved in the questionable decisions and the inadequate monitoring that contributed to the blowout. If past cases are a guide, however, the Justice Department will not prosecute rig workers who carried out decisions by their supervisors, unless they made false statements or obstructed justice, because those individuals are needed as witnesses. The Justice Department may look more closely at the role of the shore-side engineers, but those engineers appear to have been merely technical advisors; they too may be more valuable as witnesses.

That leaves only the supervisors on the rig as potential defendants, unless the Justice Department develops evidence that corporate executives were directing their activities.[19]

Media diligently covered the groundswell of support for bringing BP criminals to justice in articles about Don and me, such as the one in *The New York Times* that came out the day the indictments were announced: "In BP Indictments, U.S. Shifts to Hold Individuals Accountable." While the article had no overt suggestion that we were guilty, a quote from a former federal prosecutor would naturally color the perceptions that readers would have about Don and me: "If senior managers cut corners, or if they make decisions that put people in harm's way, then the criminal law is appropriate."[20]

The emotion behind the call of victims' family members for justice probably fueled interest in the indictments most of all. The date the indictments were announcement, *USA Today* ran these paragraphs as part of its story:

> Keith Jones, father of Gordon Jones, who died aboard the Deepwater Horizon during the blowout, said news of criminal charges was welcomed by the families of those killed aboard the rig.
>
> "For 2-1/2 years, we've wondered when anyone was going to be brought to justice for what happened," Jones said. "If in fact this involves somebody being charged or taking responsibility for the blowout itself, then I think we see it as progress."
>
> He said he expected to see multiple people charged and punished in the case.[21]

Don and I became the face of the tragedy, allowing the public to believe justice was being served.

Step 9: Closed-door bargaining shaped legal outcomes.

In BP's Third Quarter Earnings report, delivered as a webcast and conference call on October 30, 2012—exactly two weeks prior to our indictments—executives chatted amiably about the company's progress in handling its *Deepwater Horizon* problem.[22] The two speakers here are Brian Gilvary, CFO, and Robert Dudley, who replaced Tony Hayward as CEO a few months after the blowout.

Gilvary: "I would like to highlight that the US Department of Justice has been conducting an investigation into the incident regarding civil and criminal laws. We are in ongoing discussions with the DOJ and other federal agencies regarding a possible settlement of these claims and whilst we are ready to settle on reasonable terms, a number of unresolved issues remain and there is significant uncertainty as to whether an agreement will ultimately be reached."

Dudley: "As Brian noted, we have said all along that we are willing to settle if we can do so on reasonable terms and this remains our position. At the same time, we continue to prepare vigorously for trial and we will continue to update you as and when appropriate."

With those words market analysts and investors learned that BP had regular contact with DOJ and was focused on "reasonable terms" for settlement. To reach such a settlement, the two parties had to meet and determine terms agreeable to both sides. Both teams came prepared to the meetings, but determined to reach different objectives. The DOJ came for political appeasement; BP came to save money.

Both teams of attorneys negotiated behind closed doors. After the secret negotiating sessions, DOJ lawyers updated Eric Holder and other senior DOJ attorneys about possible terms of a deal; BP lawyers likely updated and informed members of the BP Board of Directors about possible terms of a deal.

Of course, I wasn't privy to any of the negotiations, but discussed probable topics with my attorneys at length. Logically, the big areas of discussion would have been DOJ demands and BP demands. Based on the outcome, which was our indictments, this is likely what happened.

On the DOJ side:

- Justice Department attorneys reminded BP that the Macondo negotiation was not going to be Texas City déjà vu. In that refinery fire, no criminal charges were filed against BP despite the loss of fifteen lives.
- The DOJ demanded that BP clean up the oil spill.
- There was a warning in 2012 that foreshadowed Deputy Attorney General Sally Yates' memo of September 9, 2015 that the Department was committed to "fully leveraging its resources

to identify culpable individuals at all levels in corporate cases."[23] The warning was that BP employees would be indicted this time. At a minimum, DOJ expected BP upper managers to be indicted.

- Justice Department attorneys asserted that BP acted with "gross negligence," which caused the oil spill. At the upcoming Federal oil spill civil trial, they would roll out their evidence and recommend the Judge impose the highest fines possible under the law.

- President Barack Obama had made commitments to the American public that justice would be served in this case and he was personally involved. In an exclusive interview on the *Today* show on June 7, 2010, Obama stated he would find out "whose ass to kick," and then kick it.

On the BP side:

- BP committed to cleaning up the entire oil spill mess.

- The citizens of the United States benefit greatly though BP's involvement in US oil exploration and production. BP employs thousands of US citizens at high paying jobs, pays hundreds of millions of dollars in US corporate taxes every year, and the US government receives hundreds of millions of dollars in oil production royalties from BP-produced oil and gas on federal lands.

- BP was not grossly negligent for the oil spill and would prove that in Federal Court.
- No BP employees were criminally negligent, and BP would fight all charges filed against BP employees.
- BP had hired superb independent attorneys, the renowned Chicago-based firm of Kirkland & Ellis. These attorneys were experienced in negotiating with the DOJ. They would pursue every path to reach a reasonable settlement.
- BP is duty-bound to uphold its own Code of Conduct, which "helps us do the right thing when we're faced with difficult decisions," according to a Letter to Employees by Bob Dudley addressing the BP Code of Conduct.

At some point in the negotiation, the DOJ had to have played its trump card: Exclude BP from competitive bidding for future oil and gas leases in the Gulf of Mexico and on all other federal offshore lease sales for years to come.

At that point BP lawyers would have returned to Sunbury England to get recommendations and possibly re-define the "reasonable terms" with the BP Board of Directors.

Step 10: BP attorneys sever communication with my attorneys.

BP paid for legal support for some of us on *Deepwater Horizon* after the disaster.

From May 2010 through September 2012, Houston-based Shaun Clarke and David Gerger were invited by BP attorneys to participate in weekly conference calls to discuss current situations, share information about other related BP cases, and discuss plans and strategies concerning overall Macondo legal cases. But in September 2012 the invitation by BP attorneys to the weekly conference calls abruptly changed.

In September 2012, my attorneys were no longer invited to participate in the weekly, Wednesday calls with BP attorneys. They seemed surprised by the sudden change, but they didn't seem acutely concerned and certainly did not speculate with me about what may have prompted it. By September 2012, I knew each attorney on our team well enough to have sensed any serious concern when BP withdrew the invitation to weekly conference calls.

I don't think anyone on my legal team realized in September 2012 that BP was in negotiation with the DOJ and offering their client as a scapegoat.

After the fact, it is easy to understand why BP attorneys stopped communicating with Shaun Clarke,

David Gerger, and the rest of my legal team. BP attorneys were discussing with DOJ about the feasibility and merits of scapegoating three BP employees—and one of the scapegoats was their client.

Step 11. With Barack Obama's November 6, 2012 re-election, the politicized lawyers at the Department of Justice could move forward with indictments of BP scapegoats.

During the first week of November 2012, negotiations between DOJ and BP were closing, a deal had been struck, and the scapegoating process had been completed by the two parties. Justice Department lawyers had written criminal indictments against BP well site leaders Don Vidrine and Robert Kaluza and against former BP vice president Dave Rainey. The indictments were sealed, awaiting the results of the November 6th general election.

Obama's re-election would give the green light to indict—to "kick-ass" as he had said he wanted to in the days after the blowout. The politicized lawyers at DOJ would unseal the three indictments against BP employees with great fanfare. It was "showtime" for them in making the announcement to news media in New Orleans of the indictments; they were heroes in the eyes of many. The BP executives and managers who knew they were accountable for the Macondo blowout also had a lot to be happy and relieved about, and so did the BP Board of Directors. BP wouldn't have to jeopardize its right to drill in the Gulf of Mexico or any other federally-controlled area.

If Mitt Romney had won the Presidential general election, a new Attorney General would have been appointed, revolving-door, politicized lawyers at the DOJ would resign until the next Democratic administration came back to power, and the career lawyers at the DOJ would have gone back to practicing law with integrity, despite having a new set of political demands.

If Mitt Romney had been elected President of the United States, the three BP scapegoats may not have been indicted.

That didn't happen, and the Obama Administration now thought it knew "whose ass to kick."

Step 12. Orders from above gave Eric Holder a green light.

Barack Obama gave the okay to unseal the indictments against the BP employees. The scapegoats had been identified and directly implicated by BP in the guilty plea agreement, which asserts that the well site leaders were negligent in supervising the tests and did not communicate with engineers on shore.[24]

Eric Holder travelled to New Orleans from Washington DC and announced the BP guilty plea criminal settlement: BP would pay huge fines and three BP individuals had been indicted—he repeated "individuals had been indicted" five times but failed to mention names of the BP well site leaders. They were simply "individuals."

There were celebrations on the top floors of the BP office buildings in Sunbury, England and Houston, Texas. Executives and senior managers got handed free passes and the US government wouldn't dare curtail BP's pursuit of oil and gas leases.

And there were celebrations in the halls of two branches of the federal government. The three BP scapegoats would satisfy the political pressure brought to bear by influential members of Congress, and deliver on Barack Obama's promise to kick someone's ass.

The big bonus was praise found in numerous media sources due to the indictments.

With final approval from the BP Board of Directors and blessing by President Obama, the final BP-DOJ criminal settlement was this in a nutshell:

- BP would plead guilty to felony manslaughter charges, environmental crimes and obstruction of Congress.
- BP would directly implicate three of its own employees for horrendous crimes. Two well site leaders would be charged with eleven counts of Seaman's Manslaughter and Involuntary Manslaughter, as well as one violation of the Clean Water Act. A former BP VP would be charged with obstruction of Congress and making a false statement to law enforcement.
- BP would pay $4 billion in fines and penalties.
- BP Executives told market analysts in

October 30, 2012 that BP was "ready to settle on reasonable terms" and nine days after Barack Obama was re-elected President, those "reasonable terms" were revealed. With just the right spin, "reasonable terms" could include the possibility of three innocent people languishing in prison for many years so BP could sustain profitability in the United States.

Step 13: BP began the process of reinventing itself.

The same day Eric Holder announced BP indictments, BP CFO Brian Gilvary held an investor conference call with stock analysts and investors. *The New York Times* reported the substance of it:

> Brian Gilvary, BP's chief financial officer, said in a conference call with analysts that the board weighed the settlement struck with the government against the prospect of a much wider criminal indictment that would have involved more people in the company. "A criminal indictment would have been a huge distraction," he said.[25]

CEO Bob Dudley reinforced the sentiment two weeks later on December 3, 2012 during BP Upstream Investor Day remarks in Sunbury:

Turning now to the start of our agenda and our US legal position. We have recently taken further steps to significantly reduce the legal risks facing the company.

This is over and above amounts already paid out of the Trust Fund or earmarked to fund the estimated $7.8 billion settlement with the Plaintiffs' Steering Committee. We can now proceed with addressing these remaining claims without the possibility of a criminal indictment and the potentially onerous consequences and distraction that would have accompanied such a proceeding.

During his November 15, 2012 announcement of the BP guilty-plea criminal settlement with the DOJ, Holder noted that the criminal investigation was ongoing. This directly contradicts Gilvary's clear suggestion on the same day in an investor conference call that the criminal investigation had ended.

Step 14: No BP board member, executive, or attorney has ever come forward to expose and renounce the unethical decisions to scapegoat fellow employees.

Challenges to the logic and ethics of indicting Don Vidrine and me surfaced as early as the day after Holder made the announcement—yet no one among BP's leadership or legal team has ever pointed to the immorality of the action.

Robert Reich was an early, strong critic of the indictments. One day after the BP-DOJ guilty-plea agreement was announced, the renowned public policy expert and veteran of three presidential administrations published the provocative "Why BP Isn't a Criminal." Dr. Reich's article is an eloquent rebuke from someone with the experience, knowledge, and guts to see through the plea agreement. Here are key excerpts:

> The Justice Department just entered into the largest criminal settlement in U.S. history with the giant oil company BP. BP plead guilty to 14 criminal counts, including manslaughter, and agreed to pay $4 billion over the next five years.

> This is loony.

> [It] defies logic to make BP itself the criminal. Corporations aren't people. They can't know right from wrong. They're incapable of criminal intent. . .

> The perfidious notion that corporations are people can lead to even more bizarre results. If corporations are people and they're headquartered in the United States, then presumably corporations

are citizens. That means they have a right to vote as well.

I'll believe corporations are people when Texas executes one.

Can we please get a grip? The only sentient beings in a corporation are the people who run them or work for them. When it comes to criminality, they're the ones who should be punished.

Punishing corporations as a whole almost always ends up harming innocent people – especially employees who lose their jobs because the corporation has to trim costs, and retirees whose savings shrink because their shares in the corporation lose value.

. . . [The] people responsible for BP's deaths and oil spill weren't BP's rank-and-file employees or its shareholders. They were the executives who turned a blind eye to safety while in pursuit of their own rising stock options, and who conspired with oil-services giant Halliburton to cut corners on deep water drilling when they knew damn well they were taking risks for the sake of fatter profits.

They're the ones who should be punished. Failure to punish them simply invites more of the same kind of criminal negligence by executives more interested in lining their pockets than protecting their workers and the environment . . .

But the Justice Department's criminal settlement with BP gives these top executives a free pass – allowing the public to believe justice has been done. Instead of going after the real criminals, the Department has gone after the schleps who got caught up in the mess. It's filed manslaughter charges against two BP rig supervisors for allegedly ignoring warning signs of the blowout that set fire to the rig, which later sank. . .

The Department's $4 billion criminal settlement with BP isn't big enough to affect the oil giant anyway. BP's market capitalization is $128 billion. Yesterday, BP's stock price closed at $40.30 a share, up 0.35 percent from the day before the settlement was announced.[26]

The scapegoating of three innocent BP employees is a depressing story about the absence of business ethics at the Board and executive levels at BP, absence of legal ethics and highest of standards of law practice at the Department of Justice, and shameful human morality exhibited by every individual that was involved. The BP Code of Conduct meant nothing, the highest standards of legal ethics which should be practiced at the DOJ meant nothing, and basic human morality was compromised for self-preservation and to appease political pressure and save money.

Some of the same politicized attorneys who participated in the Macondo injustices are still employed at the DOJ and eight unethical, immoral and unscrupulous BP board members and executives who agreed to scapegoat three innocent fellow BP employees in 2012 are still there as of this writing.

CHAPTER EIGHT

1 CORRUPT DEAL

My definition of a corrupt deal is one predicated on falsehoods and veiled agendas—in this case, political and corporate agendas. "Corrupt deal" is therefore an appropriate way to describe the guilty plea agreement between BP and the US Department of Justice in the aftermath of the *Deepwater Horizon* tragedy. This chapter explores the validity of that statement.

A methodical investigation into an accident is the path to identifying the people, things, events in time, and/or places that had a consequential role in the accident. In other words, a company cannot claim "pilot error" if investigators determine the engines cut out due to poor maintenance.

On the other hand, a shoddy investigation opens the door to cover-ups and collusion.

The National Transportation Safety Board (NTSB), an independent agency that routinely handles investigations that include maritime and pipeline accidents, had no direct involvement in the blowout investigation—and that's a shame, in my opinion. The group has been praised many times for the efficiency and thoroughness of its process, with a team from the Massachusetts Institute of Technology (MIT) even proposing in November 2010 "NTSB as a model for addressing systemic risk in industries and contexts other than transportation."[1]

The MIT team cited "impartiality and singular focus" as critical to NTSB's success: "With the primary focus on accident investigation, determination of probable cause, and formulation of safety recommendations, the NTSB is relatively free of ongoing economic and political influences."[2]

In contrast, the investigative teams first coming out with reports about *Deepwater Horizon* had distinct "economic and political influences." As explored in previous chapters, the BP and Transocean reports were configured to protect company interests. The Report to the President and Chief Counsel's Report contained evidence supporting the President's expressed desire to "kick ass." The sole investigating organization that displayed both objectivity and expertise was the United States Coast Guard.

An NTSB "Go Team," composed of technical experts who are dispatched promptly after an accident, would likely have discovered quickly that the blowout prevention equipment on *Deepwater Horizon* did not comply with the Code of Federal Regulations for four years. That lack of maintenance led directly to the blowout. Additionally, investigators would have discovered early on that the US Mineral Management Service (MMS) did not fully enforce federal drilling codes that govern offshore drilling the Gulf of Mexico for four years. Theirs is an investigative process that deserves scrutiny because it points to how one corrupt deal between BP and the Department of Justice might have been averted.

In glancing over the way the NTSB Go Team operates, there is much to learn about how methodical, thorough investigations proceed. Most *Deepwater Horizon* reports did not have comprehensive coverage of all the types of information that NTSB has on its standard checklist.[3]

The Go Team's immediate boss is the Investigator-in-Charge (IIC), a senior investigator with years of NTSB and industry experience. Each investigator is a specialist responsible for a clearly defined portion of the accident investigation. In aviation, these specialties and their responsibilities are:

OPERATIONS: The history of the accident flight and crewmembers' duties for as many days prior to the crash as appears relevant.

STRUCTURES: Documentation of the airframe wreckage and the accident scene, including calculation of impact angles to help determine the plane's pre-impact course and attitude.

POWERPLANTS: Examination of engines (and propellers) and engine accessories.

SYSTEMS: Study of components of the plane's hydraulic, electrical, pneumatic and associated systems, together with instruments and elements of the flight control system.

AIR TRAFFIC CONTROL: Reconstruction of the air traffic services given the plane, including acquisition of ATC radar data and transcripts of controller-pilot radio transmissions.

WEATHER: Gathering of all pertinent weather data from the National Weather Service, and sometimes from local TV stations, for a broad area around the accident scene.

HUMAN PERFORMANCE: Study of crew performance and all before-the-accident factors that might be involved in human error, including fatigue, medication, alcohol, drugs, medical histories, training, workload, equipment design and work environment.

SURVIVAL FACTORS: Documentation of impact forces and injuries, evacuation, community emergency planning and all crash-fire-rescue efforts.[4]

Any objective report would have had a *Deepwater Horizon* "go team" that included experts in the areas listed above. Of note is what they might have discovered—and considered relevant and important in these particular areas:

OPERATIONS: The core team conducting negative tests had more than 100 years of combined drilling experience and unimpeachable reputations in safety. These are people who wouldn't go to bed if they thought there was the hint of a problem.

STRUCTURES: Evaluation of equipment conditions prior to the event and post event would have necessitated two key things: (1) reviewing inspection and maintenance records for rig safety equipment and (2) conducting a post-mortem after the BOP stack was pulled from the water.[5] Given that the BOP had not gone through the inspection and maintenance required by the Code of Federal Regulations since its installation in December 2000, investigators would have to rely on ten-year-old information about many parts of the BOP stack. That fact alone would have alerted investigators to the non-CFR compliance issue.

POWERPLANTS: Examination of all equipment related to engaging and sustaining electrical power would have focused attention on the blue and yellow pods in the BOP stack. It was their "job" to activate the fail-safe AMF/deadman. Why didn't either one do its job? And more than examining the equipment itself, an expert in this arena would want to know about policy and practice that affected power sources. Transocean had a policy of changing control pod batteries every year. Cameron, the manufacturer of the BOP, also recommended that control pod batteries be replaced after a year of use. Despite its own policy and the manufacturer's recommendation, however, Transocean had not replaced the blue pod's 27-volt battery since November 2007.

SYSTEMS: Study of components of the rig's systems might have yielded a fact that later surfaced in my trial: Transocean violated federal regulations regarding routing the diverter to a mud gas separator rather than overboard.

COMMUNICATION CONTROL: Reconstruction of onshore-to-rig transmissions would have immediately made it impossible to assert that well site leaders did not communicate with a BP senior onshore engineer the night of April 20, 2010.

None of this is meant to claim that an organization following NTSB investigation protocols would have solved all the mysteries of *Deepwater Horizon* swiftly. It is meant to say simply that this kind of system would likely have made it impossible to indict the BP well site leaders on the rig. Additionally, it would have provided a sturdy framework of facts for the Department of Justice attorneys constructing a plea agreement with BP.

Fill-in-the-Blank Reporting

None of the investigative teams could piece together a complete story because each lacked input from key people on the rig. The teams didn't have all the facts, so they tried to construct theories about the Macondo blowout. In the cases of BP and Transocean, the investigative teams had only to look to their senior leadership to find justification for filling in the blanks in a way that benefitted their companies.

At a US Senate hearing held in Washington DC on May 11-12, 2010, Transocean CEO Steven Newman flatly dismissed the importance of the BOPs in preventing the blowout. He told members of Congress that blaming the BOPs "simply makes no sense" because there was "no reason to believe" the BOPs were not operational.[6] On the following day, he said, "The ineffectiveness of the BOP to control the flow is not the root cause of the event [blowout]."[7] When I watched the video and listened to Newman's statement it reminded me of 1994 when seven cigarette companies' CEOs stated under oath, one at a time, that they did not believe nicotine was addictive.[8]

In all fairness to members of Congress, they had only cursory knowledge about the complex exercise of offshore deepwater drilling. That first hearing was only three weeks after the blowout, so their staffs would have rushed into meetings with experts and struggled mightily to brief their bosses. Members of Congress could not focus their questioning enough to enable them to identify the cause of the Macondo blowout or to suspect that senior managers at BP and at Transocean had made deliberate decisions that broke federal regulatory law for over four years. The root cause of the Macondo blowout was that the BOPs failed, and the BOPs failed because they were not properly maintained due to negligent decisions made by Transocean and BP senior managers. Both parts of that fact would not become part of public record until years later.

BP CEO Tony Hayward set a different kind of wayward example to his team in his Congressional testimony: Say enough to suggest an answer, but don't really answer pointed questions. Hayward threw out the possibilities for the cause of the blowout that he and his legal team had targeted as acceptable: "cementing of the completed well, the casing system surrounding the well bore, the pressure testing and the procedures to detect explosive gas in the well."[9]

Members of Congress were suspicious, even without a firm grasp of the science. They detected a cover-up, but as I've said many times, offshore drilling is a complicated business and they just did not know how to pursue a line of questioning that would drive Hayward into an answer trap. The best they could do is help push him off the cliff edge that he had wandered toward already. Hayward was forced to resign as CEO six weeks later.

The exercise of avoiding critical issues or fanciful explanations for what caused the blowout went beyond Congressional testimony and investigative reports. It also found its way into courtrooms. Numerous hearings and several trials were held over a span of close to six years as a direct result of the Macondo blowout. Hearings at the Federal level and State levels were held to try to identify root causes of the Macondo blowout as well as accountabilities for the disaster and resultant oil spill. The civil and criminal trials resulted in varied rulings and verdicts. Large numbers of attorneys were hired due to the extensive number of people involved and the complexity of the Macondo legal proceedings. These many attorney teams, in turn, hired well-compensated experts to testify at trials and hearings in support of their attorneys' side of the story.

Divergent Views of "Experts"

Ever since a 1993 Supreme Court decision in *Daubert v. Merrell Dow Pharmaceuticals, Inc.,* federal trial judges are supposed to uphold standards in accepting someone as an expert witness. They are tasked with protecting juries from bogus scientific theories concocted to serve the needs of a high-paying client. In the case of the Macondo blowout, many of the supposed experts hired to testify probably met the criteria of the judges. Nonetheless, what came out of the mouths of some of them seemed like tortured speculation, at best, to those of us with the experience and specific education that make us more qualified experts than they were in deepwater offshore drilling.

One illustration of how gaps in knowledge were filled in with opinion concerns various expert testimonies on the subject of blowout prevention equipment. One would think that a group of experts would agree with each other a bit more than these people did.

On December 15, 2010, The US Department of Justice filed a civil and criminal suit against BP and its partners in the Macondo oil well, including Transocean Ltd. and Halliburton Inc., for violations under the Clean Water Act in US District Court for the Eastern District of Louisiana. The plaintiffs of the suit included Gulf States and private individuals. The DOJ sought stiff fines, and DOJ lawyers stated that the department would seek to prove BP's gross negligence and deliberate misconduct in causing the Macondo disaster. BP replied that they would prove otherwise. This standoff was the beginning of the much-publicized Federal Oil Spill trial held in New Orleans that began in March 2013 and finally ended in February 2015. Both sides hired experts to testify at the two-year trial.

When the topic of *Deepwater Horizon*'s blowout prevention equipment took center stage, these BOP experts either submitted opinion papers or testified in person at the oil spill trial:

- for the DOJ: Rory Davis
- for BP: Arthur Zatarain
- for Transocean: Greg Childs
- for Halliburton: Glen Stevick

I read all available testimony and expert papers and can say this: sometimes one or more of the BOP experts agreed with each another, occasionally most of the experts agreed with each other, and never was there complete agreement among the BOP experts.

These excerpts from the Findings of Fact[10] from the Federal Oil Spill Trial highlight why juries might wonder who really knew his stuff when it came to blowout preventers:

395. The blue pod failed to activate the BSRs [blind shear rams] on April 20 because its 27-volt battery was too weak to open the solenoid valve.

- Three of four BOP experts agreed with this Finding of Fact

397. The yellow pod failed to activate the BSRs at AMF [automatic mode function] time because its solenoid valve—or more specifically, one of the coils within the solenoid valve—was reverse wired.

- Three of four BOP experts agreed with this Finding of Fact

401. Mr. Childs, Transocean's expert, opined that the 27-volt battery in the blue pod was sufficiently charged on April 20, 2010, and that the solenoid in the yellow pod operated despite being reverse wired. The Court finds these explanations lack credibility and are unpersuasive.

- The Transocean BOP expert, Greg Childs disagreed with all other BOP experts. Transocean did not want to be held directly responsible for causing the Macondo blowout because they failed to properly maintain the BOPs, so they found a BOP expert willing to state under oath that the BOP AMF/deadman functioned properly. That defies logic, as the court determined. All power at the drilling rig was lost. The AMF/deadman is designed to activate via batteries when all other power is lost—it is failsafe—but it did not seal the Macondo well kick because no power came from the batteries.

Baseless Assertions of "Experts"

The *dramatis personae* of the trials include a few experts who might be classified as "villains"—at least by people who know a lot about offshore drilling. I've singled out some of the biggest offenders to show how much easier it is for a company like BP to forge a deal with the government when fact and fiction become intermingled.

Richard Heenan

At the Federal Oil Spill trial, the Department of Justice hired Richard Heenan, a Canadian mechanical engineer, as a drilling expert. Heenan made several comments during his testimony concerning the negative testing conducted on *Deepwater Horizon* although his resume at the time suggested he had never planned, engineered or supervised a negative test on a deepwater drilling rig.

During testimony, Heenan said results showing pressure in the well had climbed to 1,400 pounds per square inch were not a good sign and should have indicated the well was unstable. *Law360* reported the damning testimony as follows:

> According to a report Heenan helped to prepare on the events leading to the blowout, BP and Transocean workers' decision to deem the negative pressure test successful "had no foundation in well control or physics." Assistant U.S. Attorney Michael Underhill asked Heenan to explain the conclusion.
>
> "I couldn't comprehend as I looked at this how these decisions were made considering the evidence that

was there and the information that was available," he told the court. "I couldn't believe based on what I saw that the people on the rig came to the conclusion that this was a successful test."[11]

As indicated in Chapter Three, science proves that the source of the 1,400-psi pressure in the drill pipe did not originate from a Macondo reservoir pressure. The simplest way to express that proof is that records available to anyone and everyone show the 1,400-psi pressure bled off at 8:00 p.m. on April 20. If the source of the 1,400-psi pressure was from any Macondo reservoir pressure the drill pipe would have never bled to zero and the drill pipe would have flowed continuously, just as the Macondo well continued to disgorge for eighty-seven days after the blowout.

Richard Heenan was a paid expert representing the DOJ who likely failed in interpreting the Macondo well pressure profile chart. Despite ostensibly having no negative test experience on an offshore rig, he was allowed to testify in court about a critical element of the case. His testimony made BP's well site leaders look bad while it took a little heat off BP.

Adam Bourgoyne

At the Federal Oil Spill trial, Adam Bourgoyne testified as an expert for BP. Bourgoyne was an advisor.

From the Findings of Fact, the court agreed with this statement made by Bourgoyne:

551: The Court agrees with the view expressed by, among others, BP's expert Dr. Bourgoyne, that the negative pressure test would have been properly interpreted had the drill crew been permitted to continue the test on the drill pipe.[12]

In fact, the first negative test was conducted on the drill pipe from 5:18 p.m. to 5:35 p.m. April 20th and the results of the first negative test confirmed well integrity.

In my opinion Bourgoyne failed to perform basic engineering due diligence before making inaccurate and incriminating statements concerning the first successful negative test on the drill pipe. If any of the experts who testified about the first negative test on the drill pipe had simply reviewed the pressure profile chart, it would have been obvious that well integrity had been established after the negative flow test.

Bourgoyne could have used information in the BP Accident Investigation Report, Appendix Q, for well-fluid displacement information. With that information it is an easy calculation to determine what the drill pipe pressure was prior to the start of the first negative test—underbalanced to Macondo reservoir pressures from 865 psi to 1,996 psi. Next, he could have simply looked at the pressure profile chart and observed pressure was bled off from the drill pipe from 1,261 psi to zero at 5:27 p.m. The drill pipe remained open for the next five minutes. The drill pipe was open to a huge underbalance pressure, so if the casing, casing collar float or hanger seals leaked oil and gas or pressure from the Macondo well, there would have been massive flow through the drill pipe. There was no flow.

Frederick "Gene" Beck

This expert witness is particularly interesting because he introduced a scientifically baseless theory at trial.

Petroleum engineer Gene Beck testified at the Federal Oil Spill trial as a well-paid expert representing Halliburton Company, the company that planned and pumped the final production casing foam cement job. Because Beck presented a theory on the witness stand not addressed in his paper previously given to the court, BP attorneys were unprepared to challenge the unexpected testimony, and therefore failed to illustrate that Beck's theory was scientifically baseless.

During testimony, Beck theorized that Halliburton employees had pumped a successful production casing cement job, but the cement had not travelled through the bottom of the casing and up the casing annulus. Instead, he theorized that the casing had burst somewhere below the casing float collar and the cement travelled through the burst opening in the casing and then up the casing annulus. He further presented a theory about the likely flow path of the oil and gas from the lowest Macondo reservoir formation to the drilling rig floor. And it was that formation, according to Beck, that was the source of the Macondo well kick. Beck has a stunning imagination.

The flow-path theory he presented in Federal court was scientifically impossible and poles apart from the flow-path theory of a BP expert, who offered an equally scientifically odd explanation of what happened. But BP trial attorneys were unprepared for these assertions; on the spot, they couldn't discredit Beck. Consequently, Gene Beck was quoted many times over as having *the* answer to the genesis of the disaster. In a way, even the *Deepwater Horizon* movie seemed to draw on this explanation by having issues with the cement foreshadowing the tragedy.

Two "impossible" parts of Beck's theory are (1) he asserted that the production casing encountered debris while it was run in the hole and was under 140,000-psi compressive force at setting depth and (2) the sudden release of 3,142-psi pressure when the float valve shifted burst production casing—casing rated to withstand over 13,000 psi of internal pressure before burst.

Those assertions border on absurdity, with the second one being an obvious gaffe. Regarding the first one, I was on the drilling rig floor when we ran the production casing in the open hole. We encountered no hole problems and did not set the casing string in any type of debris. In fact, there was 56 feet of open hole—called a "rat hole"—below the production casing setting depth.

A Plea Agreement Built on Deceit

Partial truths, critical omissions, and even deliberate lies permeated BP's guilty-plea agreement, just as they had the eight investigative reports as well as expert testimony before Congress. On November 15, 2012, DOJ had a plea agreement in hand signed by David J. Jackson, BP's Company Secretary and Mary Jane Stricker, Assistant Corporate Secretary. Multiple false statements, which their signatures affirmed to be true, were the basis for the indictments of Dave Rainey, Don Vidrine, and me. Officially, Don and I were indicted on November 14, 2012, but that was not made public until a day later.

Based on what has already been discussed throughout this book, many statements in the plea agreement would strike you as shocking. They are all preceded by the sentence, "Defendant BP Exploration & Production, Inc ("BP") agrees that, if the case were to proceed to trial, the Government could establish beyond a reasonable doubt that . . ." With that statement, it is now possible to see the culmination of bad testimony by expert witnesses and flawed investigative reports from multiple sources.

I've underlined inaccuracies in the following short section of the plea agreement to call attention to a few things the Government said it "could establish beyond a reasonable doubt." All are taken from Exhibit A[13] of the plea agreement.

On the night of the explosion, BP had two Well Site Leaders on the *Deepwater Horizon*, who were BP's employees, agents, and highest-raking representatives on the rig. The Well Site Leaders were responsible for supervising the negative pressure test conducted by Transocean. On or about April 20, 2010, between approximately 5:00 and 8:00 p.m. Central Daylight Time, the negative pressure test performed on the Macondo Well provided multiple indications that the wellbore was not secure. BP's Well Site Leaders negligently supervised the negative pressure test during this time, failed to alert engineers on the shore of these indications, and along with others, ultimately deemed the negative pressure test a success, all in violation of the applicable duty of care. The negligent conduct of BP's Well Site Leaders is attributable to BP.

To clarify, point by point, why the underlined statements are false:

- Normally, Don and I would have been the highest-ranking BP representatives on board, but Pat O'Bryan, Vice President of Drilling and Completions, and David Sims, Drilling & Completions Operations Manager, happened to be on the rig on April 20. In an official federal government document—one of huge significance in terms of fines levied on BP— one would think that each statement would be checked for correctness. This "slip" is indicative of how shoddy DOJ's work was in the plea agreement.

- The agreement refers repeatedly to a single negative test; there were two. Throughout this section, every reference should be to negative tests.

- The negative tests gave no indication that the wellbore was not secure.

- We could not have negligently supervised the test since there were no guidelines or standards for the test. An important fact is that there are no industry standards for negative testing, no standard policy or procedure for negative testing at BP or Transocean, and no formal employee training for negative testing. Add to that the fact that the BP wells team provided no detailed negative testing procedure to follow, and there is an airtight legal argument that we could not have "negligently supervised the test." More importantly, our experience with doing flow tests and our interactions with highly experienced Transocean personnel during the tests would suggest we were diligent, not negligent.

- The BP investigation team knew there was a rig-to-shore telephone call at 8:52 p.m. during which the BP well site leader did alert the BP senior operations engineer in Houston on

April 20, 2010. The DOJ also knew about that telephone call and substance of the conversation yet allowed the lie in the plea agreement.

- We could not violate the applicable duty of care because there was no standard of care and because both negative flow tests were successful.

A Financial Slap on the Wrist

Headlines in mid-November 2012 announced the "big money" deal associated with BP's guilty plea agreement. The Department of Justice had scored fines and penalties in the billions. On the heels of Barack Obama's re-election and leading promptly to the indictments of BP's *Deepwater Horizon* well site leaders, the agreement helped generate a maelstrom of celebrations for many Americans.

To most people, "billions" sounds like a spectacular sum of money, sure to eviscerate a company's financial health and to aggravate shareholders. BP's senior executives and board members are not most people, though. They look at profit and loss statements with numbers in the billions all the time. The terms in the plea agreement would not have made them smile, of course, but they would not have kept them awake at night, either.

I don't want to minimize the impact of BP's payments related to the tragedy on the company and its shareholders, but rather look at how profoundly DOJ's perceived victory distracted from further investigation of the real cause of the blowout and where responsibility for it should rest. The triumph of a record-setting financial settlement satisfied DOJ's political aims and protected BP's corporate agenda. The plea agreement and subsequent settlements had nothing to do with *real* justice. Real justice would have taken aim at the true villains in this disaster.

Fines, Penalties, and Settlements

The terms of BP's guilty plea agreement cover the following:

- Payment of criminal recoveries totaled $4 billion.
- Payment of criminal fines totaled $1.256 billion.
  - Each of the eleven counts of Seaman's Manslaughter to which BP pleaded guilty carried a fine of $500,000.
  - Violation of the Clean Water Act carried a fine of $200,000, or $25,000 per day, with the ultimate amount paid to the Oil Spill Liability Trust Fund established at $1.5 billion.
  - Violation of the Migratory Bird Treaty carried a base fine of $15,000, but the final penalty was $100 million to be paid to the North American Wetlands Conservation

Fund. The purpose of the Fund is "restoration of wetlands and conservations projects located in States bordering the Gulf of Mexico designed to benefit migratory bird species and other wildlife and habitat affected by the Macondo oil spill."[14]

o Obstruction of Congress carried a fine of $500,000.

Credit card companies are not the only entities with payment plans; so is the federal government in some cases. Built right into the plea agreement are payment terms. The $4 billion penalty would be paid out over five years. Other details are:

- The $5.5 million related to Seaman's Manslaughter had to be paid within sixty days of sentencing. The same terms applied to the $500,000 BP owed for "Congressional Obstruction."

- The $1.6 billion related to violations of the Clean Water Act and Migratory Bird Treaty had a much more extended payment plan: $250 million to be paid within sixty days of sentencing; the next $250 million to be paid within one year of sentencing; another $250 million to be paid within two years of sentencing; another $150 million to be paid within three years of sentencing; another $150 million to be paid within four years of sentencing; and the remainder to be paid within five years of sentencing.

Beyond the Agreement, Good News and Irony for BP

BP's financial obligations were not limited to the amounts in the plea agreement. There were additional costs of a $20.8 billion settlement related to cleaning up more than 1,300 miles of Gulf coastline as well as other settlements that kept adding to BP's bill.

When all the numbers were in, they totaled $61.6 billion that BP owed to a variety of entities. At first, that seems an almost incomprehensibly large amount, but a close look at how it breaks down makes the amount seem like a cup of water in a bucket.

Good News

First, $61.6 billion is a pre-tax number. After the tax impact is factored out, the hit to BP is $44 billion. This is a combination of settlements with federal and state authorities, property owners in the Gulf region, and individuals including shareholders.

Second, the bulk of these payments will be made over the course of *sixteen years*, with the clock starting in 2016.

Irony

Despite having managed itself out of a sizable financial burden, BP has never acknowledged the financial irony related to the disaster. A quick review of the numbers shows that the cost of the *Deepwater Horizon* tragedy was roughly *1,000 times as much* as the company saved by not maintaining the BOPs.

My first, very conservative estimate of how much BP saved by not pulling the BOP stack out of the water for repairs was $15 million, which was based on an estimated per/day operating cost of $1 million and minimum downtime of fifteen days. For a complete inspection and repairs to be done in compliance with federal regulations, the MODU *Deepwater Horizon* would have had to be sailed forty miles into dry dock to have larger cranes at the dock handle the BOP. Time in dry dock would range from sixty to ninety days, so BP could incur $60 to $90 million in operating costs and Transocean could lose $30 million to $45 million in revenue stream.

I have heard the question several times: Were BP's negligent actions to save money typical in the industry? Obviously, there is no way for me to know that for sure, but two points help suggest an answer.

First, the importance of maintaining blowout prevention equipment has been engrained in industry practices for decades. Blowout preventers have been standard safety equipment in the drilling industry since shortly after they were introduced by Cameron, the manufacturer, in 1922. Cameron's invention is credited with revolutionizing safer drilling operations in an industry that was in its infancy in the early part of the twentieth century. The American Petroleum Institute (API)—the leading force behind oil and gas industry standards for more than ninety years—published its first set of recommended practices for maintaining blowout prevention equipment systems for drilling wells in February 1976. As is common for the industry, API's recommended practices later become the mandates of the Code of Federal Regulations. In other words, the value and necessity of maintaining blowout preventers have been recognized by the industry for decades and the requirement to do so by the federal government followed.

Second, we can look to the recorded instances of successful prevention of blowouts as indicators that compliance is more common than non-compliance. According to David Uhlmann, the former chief of the Justice Department's environmental crimes section:

In the United States, there were twenty-eight major drilling related spills, natural-gas releases, or well-control incidents in the Gulf of Mexico during 2009, including a loss of well control and an explosion in April that did not result in a major spill only because the blowout preventer worked. No criminal charges were filed.[15]

In summary, the cost to BP of complying with the Code of Federal Regulations would have been about $60 million. The financial obligation resulting from the *Deepwater Horizon* tragedy was more than $60 billion.

The Rewards of Deceit

The financial impact of BP's plea agreement is minimal compared to what BP received in return. The rewards of the leniency deal are sizable and more significant to the company in the long-term: protection of the company's earnings potential and shielding of its senior executives and board members from prosecution. Those rewards could only come from what I've termed a "corrupt deal."

In discussing the aftermath of the financial crisis of 2008 in his book, *The Chickenshit Club,* Jesse Eisinger describes how little fines and prosecutorial threats from the Department of Justice affected organizations like Bank of America and JP Morgan. The same words could be applied to the post-deal world of BP:

The government celebrated its admissions of wrongdoing and guilty pleas, but they are little more than a semantic change. Prosecutors took extreme measures to minimize the regulatory consequences for a guilty plea. Regulators did not pull licenses . . . The guilty pleas had only symbolic value. They lacked force just as much as the old settlements did.[16]

CHAPTER NINE

1 MAN STANDING

A mong the three scapegoated BP employees, I was the last one to find resolution.

The case against Dave Rainey ended on June 5, 2015 after a twelve-person trial jury deliberated for less than two hours before finding him not guilty of all crimes. Don Vidrine was in very poor health and told his legal team, "I can't take it anymore." He went for a misdemeanor plea bargain on the Clean Water Act charge.

I was determined to prove my innocence of all wrongdoing and my attorneys at the Gerger & Clarke law firm were resolute in ensuring the same outcome. In February 2016, I finally got my days in court. The evidence presented at my trial was so compelling that it took the jury less than two hours to acquit me of all charges. Even before the trial, though, a great deal happened to allow me to be the last man standing among those accused of heinous crimes.

Fighting Manslaughter Charges

In one-inch tall letters, the word FELONY is stamped on the cover of the 21-page indictment reading "United States of America v. Robert Kaluza and Donald Vidrine." The title is "Superseding Indictment for Involuntary Manslaughter, Seaman's Manslaughter and Clean Water Act."

I had been a productive employee of a major international company doing a job that I had done safely for thirty years when the tragedy occurred on *Deepwater Horizon.* The word "felony" applied to things criminals did, not to anything I did. Those inch-tall letters made me furious and I immediately committed to proving my accuser—the United States Department of Justice— wrong in taking this action against me.

As I read through it, I came to the list of the eleven people who had been killed on the rig. The next sentence was the most painful, aggravating, and invalid statement in the entire document: "The negligent and grossly-negligent conduct of defendants KALUZA and VIDRINE proximately caused the deaths of these eleven men."[1]

The Department of Justice had taken on the task of proving some outrageous claims—like that one—were true beyond a shadow of a doubt. Knowing what really happened, I didn't believe they would be able to do it. At the same time, I had no idea what a convoluted and protracted process had just begun. I would have no semblance of a normal life for the next few years and my reputation would suffer. Family and friends would be reading articles effectively saying my actions had caused death and destruction in the Gulf of Mexico.

Seaman's Manslaughter

Title 18 Section 1115 of the U.S. Criminal Code opens with this paragraph:

Every captain, engineer, pilot, or other person employed on any steamboat or vessel, by whose misconduct, negligence, or inattention to his duties on such vessel the life of any person is destroyed, and every owner, charterer, inspector, or other public officer, through whose fraud, neglect, connivance, misconduct, or violation of law the life of any person is destroyed, shall be fined under this title or imprisoned not more than ten years, or both.[2]

The Seaman's Manslaughter statute had its roots in legislation of 1838, despite the reluctance of Congress at the time to regulate private industry. Thousands of people had been killed throughout the early 1800s in steamboat accidents, but it wasn't until United States Senator Josiah Johnson died in a steamboat explosion that President Andrew Jackson called on Congress to take action. Five years later, the statue became law, and according to Jeanne Grasso, an attorney specializing in maritime and environmental law, "The 1838 Act was the beginning of federal regulation of the private sector for public welfare reasons and set the precedent for the bevy of consumer protection legislation enacted during the 20th century."[3]

Despite the new law, the accidents continued, and by 1852, more than 7,000 people had died in steamboat accidents. That year, President Millard Fillmore signed a tougher bill into law. Inspections of boats increased dramatically, and the country saw a rapid drop in the death rate.

One of the most sensational prosecutions under the Seaman's Manslaughter statute bears a resemblance to the *Deepwater Horizon* case in fundamental ways. When the *General Slocum* burst into flames on New York's East River in 1904, more than 1,000 people died. A significant problem that added to the fatalities was poorly maintained safety equipment. After the fact, it was also discovered that the company that owned the boat had falsified records about the condition of the vessel. In addition, there seemed to be some rubberstamping on the part of inspectors: One month prior to the fire, the vessel had passed a safety inspection by the US Steamboat Inspection Service. Perhaps a well-read DOJ attorney saw history repeating itself in some ways with the *Deepwater Horizon* disaster and introduced the potential value of the Seaman's Manslaughter statute.

The statute languished in obscurity for quite a while, with "roughly eight major prosecutions, spanning 1848 through 1990."[4] After that, it came to life, with six prosecutions under the Seaman's Manslaughter Statute in the late 1990s and early 2000s.

Considering I had no maritime responsibilities on board the *Deepwater Horizon*, it seemed unreasonable that a law like this would apply to me. Yet looming ahead was the remote possibility I would be sentenced to 110 years in prison on the Seaman's Manslaughter charges alone.

In the statute, it soon becomes clear the applicability extends to people with responsibility for the condition of the vessel, including corporate management. Ultimately, this is how the DOJ was likely able to charge BP with Seaman's Manslaughter—a charge that BP didn't mind at all because it implies simple negligence rather than gross negligence or malice.

The charges of Seaman's Manslaughter stuck with me for only a little more than a year. My attorney team at Gerger & Clarke effectively built the case that DOJ was contorting the intent of the statute by rolling these charges into the indictment.

On December 10, 2013 Federal Judge Stanwood Duval dismissed the Seaman's Manslaughter charges against Don Vidrine and me. On March 21, 2014, DOJ appealed Judge Duval's Ruling, but on March 11, 2015, the 5th Circuit Appellate Court upheld unanimously Duval's ruling of dismissal. Duval was clear that DOJ's attempt to use Seaman's Manslaughter provisions against BP well site leaders was a stretch. He said, "...we are left with...

the incongruity of applying a statute originally developed to prevent steamboat explosions and collisions on inland water to offshore oil and gas operations—all approaching a bridge too far."[5]

Involuntary Manslaughter

Each count of involuntary manslaughter carried with it a potential prison sentence of up to eight years. With the indictment citing both seaman's and involuntary manslaughter, therefore, I was looking at a worst-case scenario of 198 years in prison before we even got to the penalty associated with violation of the Clean Water Act.

After the November 15, 2012 indictment, however, some of the information that DOJ felt was solid began to crumble. The kind of gross negligence and serious recklessness associated with involuntary manslaughter could only be proven if Don Vidrine and I had (1) failed to administer the negative tests properly and (2) failed to contact onshore engineers about our readings from the tests.

Late in 2015, the Department of Justice finally faced the inevitility that the government would lose an involuntary manslaughter case against us. Although DOJ had denied knowing about Don's 8:52 p.m. phone exchange with BP's onshore Senior Drilling Engineer Mark Hafle—as evidenced by the language in the BP plea agreement and in our indictment—the documentation of it existed. Combined with proof of test results and mounting scientific evidence that the "botched negative test" theory was flawed, that phone-call documentation pushed DOJ attorneys onto fragile branches that would soon break.

In a statement on December 2, 2015, DOJ spokesman Peter Carr admitted that the department had to pull back from the involuntary manslaughter charges:[6]

> The government moved to dismiss the remaining counts because circumstances surrounding the case have changed since it was originally charged, and after a careful review the department determined it can no longer meet the legal standard for instituting the involuntary manslaughter charges.

In the aftermath of the government's action, voices of both reason and outrage surfaced in the media. There were those who wanted to see two people from BP— probably *any* two people—go to prison for causing loss of life through their gross negligence. And then there were people who tried to put the tragedy and the crimes into perspective. Some even tried to target the true villains. One of them was David Uhlmann, a University of Michigan law professor and former chief of the Justice Department's environmental crimes section. Uhlmann pointed the finger at BP executives:

The problem in the Gulf oil spill is not that the government tried to hold individuals responsible. The problem is that the responsibility lies with the senior corporate management that created a corporate culture that promoted risk taking and did not place sufficient emphasis on safety and environmental protection.[7]

My Days in Court

In a way, my trial began with presentations about *Deepwater Horizon* that are like the material I provided in early chapters of this book. My attorneys engaged the jury in information about the operation and purpose of the rig. After familiarizing them with the *Deepwater Horizon*'s mission and function, they moved toward an explanation of the number and type of redundant safety systems that were part of the rig. Ultimately, they demonstrated that the central issue was this: Every component of the blowout prevention equipment failed. Nothing else—*nothing else*—explained why this tragedy occurred.

My lead trial attorney, Shaun Clarke, introduced me in a way that humanized the experience from my perspective. Then over the next week, a combination of my attorneys' presentations, riveting testimony from experts, and animation illustrating the functioning and malfunctioning of safety equipment on the rig led to a swift verdict. Shaun Clarke's succinct introduction was:

CLARKE: Good morning, ladies and gentlemen. Thank you for your time and attention.

You are going to be making an important choice at the end of this case.

The only thing I agreed with that the prosecutor said was right at the outset when she said, this is a case about choices. That is absolutely correct.

Lots of choices made, but Bob Kaluza didn't make the choices that led the *Deepwater Horizon* to doom.

At 9:48 p.m. on April 20, 2010, the *Deepwater Horizon* looked like this. A massive drilling vessel floating on calm seas. And as the clock ticked toward 9:49, below deck in Room 310, Bob Kaluza had been asleep for two hours. And in that final moment of peaceful sleep, he had no idea that he was awake – about to be awakened by an explosion, or that the night of April 20th would morph into a legal nightmare in which six years later he finally gets his day in Court.

At 9:49, Bob heard the explosion. He flipped the light switch, but there was no power. He fumbled in the dark, found his hard hat, goggles, and steel-toed boots, put them on, and stepped into the dark hallway. He turned toward the exit, but the path was blocked by debris from the explosion. He turned the other way, walked, and found a doorway. He didn't know exactly where to go.

And at that moment, a man or an angel came to Bob with a flashlight and led him to a lifeboat. Bob still doesn't know who that man was, but because Bob made it home safely, he is the defendant here today.

Bob Kaluza was a four-day substitute on the *Deepwater Horizon*, and the Macondo well was under control for every single second of his watch.

When Bob's watch ended at 6:00 p.m., he could have gone to bed. He'd worked 12 hours. But he stayed in the drill shack to observe, to assist if he had been needed, and to learn from a crew that he had been told was one of the very best in the business.

At 7:50 p.m., two hours after Bob's watch ended, the regular on duty, well site leader, Donald Vidrine, declared the negative test successful.

Vidrine was no rookie. He had 39 years of experience; 25 of those as an offshore well site leader. And the experienced men on the Horizon night crew, Jason Anderson, Dewey Revette, Don Clark, and Steve Curtis all agreed with

Vidrine that the test was good. That is no surprise. Because Vidrine and the crew with Bob present had just followed the written instructions they got from BP's onshore engineers to the letter.

Bob Kaluza, the four-day substitute, didn't jump up and say, "Hey, don't follow your instructions. I know better than all of you." And for that, he stands accused of perpetrating the greatest environmental crime in our nation's history.

He is innocent.[8]

Over the course of the previous chapters, I've given you essentially the same information that my attorneys provided during my trial—from who on the rig was responsible for what job to the specifics of the successful negative tests. Rather than repeat myself, I will cut to the witness and the moments in his testimony when jurors opened their eyes like they were getting birthday surprises. This truly expert witness gave them a series of plain-English explanations that were revelations about the true cause of the *Deepwater Horizon* tragedy.

The turning point was the testimony of Charlie Jones, whose court appearance focused specifically on the blowout prevention equipment.

No one had heard the story this way before. No record stating a definitive cause of the blowout had appeared in reports or previous testimony. Charlie Jones' testimony at my trial on February 24, 2016 drew a line in sand. If you believed him, I would be acquitted. If you doubted him, then the outcome of deliberation was uncertain.

Jones had experience with blowout preventers on offshore rigs like no one else I have encountered. My legal team did many extraordinary things and finding Charlie Jones was one of them.

Jones told the court he had taken a job in an oil patch—a folksy way of referring to an oil-production area—right after high school and then began working on offshore drilling rigs. After spending years installing, testing, commissioning, and repairing blowout preventer equipment for the Hydril Company—the company that invented the annular preventer in the 1930s—he rose to the top executive positions as Forum Oilfield Technologies and Forum Energy Technologies. He had a down-home style of explaining things, and that helped the jury find some complex concepts easier to digest.

The questioning at this phase of the trial was done by my attorney Dane Ball, who invited Charlie Jones to dive deep into the ocean and describe conditions where the *Deepwater Horizon* BOP stack lived.

> JONES: The way I've described it to people in the past, it's like being on the moon. In fact, it's like being on the far side of the moon. It's very cold. Ocean temperatures, roughly, at that depth, are very stable. They're just above freezing, somewhere below 40 degrees on average. So you get an idea, it's very cold.

> You couldn't see this. There's no light down there. So you're in pitch black conditions, and it's very, very cold, and you can't get down to reach it. This equipment is remotely controlled. So it's a very hostile, forbidding environment.

> And, of course, you got the pressure of the ocean at that depth working on it as well.

BALL: So before we move off of this, and we'll come back to it, could you talk a little bit about preventive maintenance and how that ties into the conditions that you've just described?

JONES: Yeah. We're nearly a mile deep. This is safety equipment. It's imperative that the safety equipment works. It works the first time, every time, and you have no recourse. It's so far away, you can't get to it. So it has many, many redundant systems embedded within its design to help you in case you have a problem. These are mechanical systems, and they're very complex, and you need to make sure they have a lot of redundancy.[9]

At that point, Jones took the jury past the shell of the BOP stack and inside to look at the pieces and parts. He described the wellbore coming up the center and went from top to bottom in identifying the components. During his explanation, the jury was able to see a visual of the BOPs, so to some extent, they could get a sense of crawling inside the gigantic assembly of safety equipment weighing 350 tons and standing 53 feet tall.

JONES: So I'm going walk you down through the stack . . . Here's the upper annular. Below it is the lower annular. This is the connector that actually connects the lower marine riser package, or LMRP, to the rest of the blowout preventer. Then we step down. This is the blind shear ram. The casing shear ram. A variable bore ram. Another variable bore ram.

BALL: And we're going to come back to those in detail, but just generally speaking, what do all of these things – what are they intended to do?

JONES: Well, each one independently is designed to shut the well in. Each one has the ability alone to shut the well in. That's what they're designed to do.[10]

Charlie Jones testified that the blowout preventer was last certified on December 13, 2000. He told the jury in no uncertain terms "this equipment is certified every five years. So this is well beyond the five-year certification requirement."[11]

He was adamant that this was a grave breach of the maintenance requirement for the most important equipment the rig had; it's known as Priority A: "...this is critical equipment, critical equipment items that may lead to loss of life, a serious injury, or environmental damage."[12]

The point was made earlier in the book—several times—that a dollars-over-safety reason why BP did not order the required inspection and maintenance is the amount of time it would take away from drilling. The price tag for cessation of operations on *Deepwater Horizon* was more than $1 million a day and the project was already behind schedule and overbudget.

Jones explained the process of extracting the BOP stack and what would happen after that, when it went it for inspection and maintenance. As a side note, multiple news sources reported that merely raising the BOP stack out of the water, nearly five months after the accident, took a crew of engineers 29 hours.[13] Before any of what Jones describes could occur, therefore, crews would have spent more than a day just getting the BOP stack to breach the water and then move it to a vessel for transport. (In Chapter Ten, I explain how structural information about the *Deepwater Horizon* BOP stack that surfaced later proved it could not even been transported in this conventional manner.)

JONES: Typically what occurs is the blowout preventer is disassembled, which generally requires the rig to be in port. So the rig is in port. The blowout preventer is disassembled. All of the big bodies, like the upper annular and the lower annular and the variable bore rams, those big items are then sent back to the OEM. So in this case, it would go back to Cameron –

BALL: I'm sorry to interrupt. OEM. What's an OEM, just in case?
JONES: Excuse   me.   Original   equipment

manufacturer. Cameron was the OEM of this equipment.

BALL: And what would happen when it got in to the OEM?

JONES: So Cameron would bring this back into their shop, and then they would completely disassemble the actual blowout preventer. So the frame and all that structure it would stay at the shipyard, but the actual pressure-containing bits, like the entire annular itself, would go back to the shop. They would take it apart. Since they are the original equipment manufacturer, they would do some very careful dimensional inspections.

This equipment – the reason dimensional inspections are important is you would think, well, things wear and so you want to check for wear. But, more importantly, this is pressure-containing equipment, so it can actually deform under pressure and you don't see it. And if it deforms, that's kind of a precursor to potentially a failure of the material.

So it's very important to do these dimensional inspections. As well, there is a lot of welding done on these pressure vessels. You have flanges and different configurations of the actual forgings that make them and you have to weld them together.

There's no way to forge them as one piece. So you do different inspection techniques, what they call surface inspection techniques where you're looking for cracks at the surface, as well as what we call volumetric inspection techniques where we're going with an x-ray or ultrasound and we're looking for failures you can't see at the surface with your eye. And those are cracks. And those cracks can propagate into failure of the device.

BALL: And what can happen if you don't do this five-year overhaul?

JONES: Oh, the equipment won't shut the well in. It'll fail.[14]

Up to this point, Charlie Jones stated in various ways that

- the BOPs are the most important safety equipment on a rig,
- each component is designed to shut-in the well,
- this equipment must be inspected and maintained every five years at a minimum, and
- if the equipment isn't maintained properly, it will fail.

The revelations kept coming for the jurors. Dane Ball probed into how usual or unusual this maintenance issue with BOPs might be.

BALL: Let me ask you: In your – how many years have you been in the BOP industry? Can you put a number on it?

JONES: 35.

BALL: In your 35 years, is this type of maintenance on a BOP something that's highly extraordinary to you?

JONES: Yes. I've never seen anything like it.

BALL: And you've never seen anything like it. Explain that. Why is this important to you?

JONES: Well, this is safety equipment. And all of my exposure has been with clients that just treat it that way, and it's the first thing that they do.[15]

Driving home the point that the BOP had redundancy upon redundancy in terms of safety equipment designed to prevent catastrophe, both my attorney and the expert witness went through the failed operation of the annulars, shear rams, and finally, the AMF/deadman. Jones had a simple conclusion regarding how likely this was to occur on an offshore rig: "It's clearly highly extraordinary. . . I've never seen it before."[16]

Dane Ball probed more about the concept of systems backing up systems. "Tell me about something called redundancy after redundancy?" he asked Charlie Jones. Jones then explained the workings of the blue and yellow pods—the power sources for the fail-safe AMF/deadman in the BOP stack—in terms anyone would understand.

> JONES: These pods are designed to be mirror images of each other. They're designed to talk to each other electronically and kind of say, "Hello. How are you? "I'm fine." "Are you okay? "Are you sure?" And they're talking all the time.
>
> But at the same time, either one of them can perform all of the functions on the BOP. So you can have failures of certain items within a given pod and still be functional on the seafloor because the other one will do it for you, you know.
>
> And these pods are very complicated. They have, I will just say, over 100 different functions that they perform. So, really, when you think about it, the odds of the same function failing on both pods at the same time, being if they're properly maintained, are very low. And that provides several layers of redundancy right there.[17]

Jones reinforced the remoteness of the equipment that is meant to protect both personnel and environment: "You know, you're down on the ocean floor, and you can't get to this to work on if it fails. So you need to be very, very proactive in your maintenance program to prevent a failure on the seafloor."[18] The clear message was this: Any company executive making the choice to forego the "bother" of BOP maintenance is a person who disregarded the safety of people onboard *Deepwater Horizon* and the health of the Gulf of Mexico.

The coup de grâce in this case is embodied in this exchange:

BALL: Let me ask you a question: If the blowout preventer had been properly maintained, was it strong enough to close in this well –

JONES: Easily.

BALL: – during the kick?

JONES: Yes.

BALL: Was the annular strong enough to close in this kick?

JONES: Yes.

BALL: Was the first variable bore ram, on its own, strong enough to close in this kick?

JONES: Yes.

BALL: The second variable bore ram?

JONES: Yes.

BALL: The AMF via the blind shear ram?

JONES: Yes.

BALL: In your 35 years, Mr. Jones, in 2010 would you have gone out to the *Deepwater Horizon* knowing what you know now about the maintenance history?

JONES: No, I would not.

BALL: Are the maintenance issues that you've talked about here today highly extraordinary?

JONES: Yes. Unprecedented, in my opinion.[19]

Stipulation No. 3

A critical document filed in the course of my trial was something known as "Stipulation No. 3." It is dated

February 22, 2016, two days before the testimony of Charlie Jones cited above. Stipulation No. 3—a long time coming—was an admission that BP and Transocean were not law-abiding entities. The short document, in its entirety, is as follows:

> The United States of America and defendant Robert Kaluza, through counsel, hereby stipulate and agree that the following facts are true and correct, and the jury may regard them as having been proved beyond a reasonable doubt:
>
> 1. The companies Transocean and BP violated federal regulations governing the
> Deepwater Horizon blowout preventer ("BOP") regarding maintaining and certifying the BOP and maintaining and certifying the "AMF/Deadman."
> 2. Transocean violated federal regulations regarding routing the diverter to a mud gas separator rather than overboard.[20]

In short, as we were into the trial, the government's case unraveled. Jennifer Saulino, Director of the Deepwater Horizon Task Force and lead DOJ attorney at my trial, had signed a document stating that Transocean and BP had made choices to violate federal regulations that resulted in loss of life and massive environmental destruction.

Verdict

It was late afternoon when the jury deliberated. In under two hours, the jury re-entered the courtroom and the deputy clerk read the following: "As to the charge set forth in the indictment in this matter, the Clean Water Act violation, we the jury unanimously find the defendant, Robert Kaluza, not guilty."

Exoneration came after 8,000 hours of legal preparation, with an estimated 90 million documents contributing to the body of information over a nearly six-year period. The result was a non-story, apparently, except to the legal community. After all those years of media people trying to reach me by phone, which I was forbidden to answer, and occasions of people snapping photos of me while I rode my bike, the Associated Press was the sole news outlet covering the verdict.

In contrast, the legal media gave Shaun Clarke and David Gerger the equivalent of a (well-deserved) standing ovation. Former prosecutor Robert Hinton went on record as saying, "This is one of the greatest legal victories against the U.S. government of the past decade. The techniques, strategies, work ethic and skills demonstrated by David and Shaun in this case should be taught in law school and in CLE [Continuing Legal Education] programs across the country."[21]

And Then What?

The title of Walter Pavlo's January 8, 2018 article in *Forbes* sums up the aftermath: "Two Years After Ruling,

BP Engineer Still Carries Burden of Prosecution." No reasonable person would assume that all the dark clouds would magically disappear after a not-guilty verdict, and that's not what I expected. I did expect an uptick in questions related to the true cause of the disaster, however. I fully expected investigative journalists and victims' families and co-workers to be vocal about the truth behind the tragedy: The BOPs didn't fall out of compliance because some mid-to-lower level employees failed to do their job. They fell out of compliance because BP and Transocean senior management decided that was an acceptable risk.

CHAPTER TEN

2 UNETHICAL GIANTS

n 1988, a *Parade* magazine reader from Evansville, Indiana asked columnist Marilyn Vos Savant: "What characteristics would you look for to evaluate a person's intellectual ability?"

She responded: "If a person knows 'what' happens, they have average ability; if they know 'how' it happens, they have superior ability; if they know 'why' it happens, they have exceptional ability." [1]

This final Chapter will enable you to earn the "exceptional" label. It is all about "why." Why did BP scapegoat these particular long-time employees? More importantly, why did the United States Government accept this perversion of justice?

Why the Well Site Leaders?

BP convinced the Department of Justice that the two of us serving as well site leaders had made mistakes with tests we conducted, and those mistakes made us culpable. On the surface that explains why fingers pointed at us, but it does not give the complete answer to why we were scapegoated, or how BP was able to carry it off.

Let's look at the "carry it off" part of this first.

There are three parts of the explanation of why BP was able to perpetrate the fraud the led to my indictment and that of Don Vidrine:

- Marketing
- Ignorance
- Politics

Marketing

By its nature, scapegoating allows people to blame others instead of accepting the consequences of their actions. It's a perennial tactic in the playground: Blame the accident on the kid standing closest to it when it happened. There was a lot more sophistication in BP's scheme to scapegoat Don Vidrine and me, however. We were doing something unrelated to the real cause of the blowout, so BP could not get away with scapegoating us unless they were also successful in covering up the truth.

BP's model for scapegoating might be described as take-off on the story of Mrs. O'Leary's cow.

For generations, people believed that Catherine O'Leary's cow started the Great Chicago Fire of 1871 by kicking a lantern over in the barn. The fire engulfed Chicago leaving 300 people dead and 17,450 buildings incinerated. Late twentieth-century forensics exonerated Mrs. O'Leary's cow by proving that a pipe-smoking

2 UNETHICAL GIANTS

neighbor of Mrs. O'Leary's started the fire by carelessly tossing a match into the barn.

The similarity is this: When people who have been affected by tragedy hear a plausible explanation for what caused it, the tendency is to tenaciously cling to it, spread it, and find evidence to support it. The first story to make sense catches the attention of those who hear it and, on some level, it satisfies them. Then it's like an earworm—a song you cannot get out of your head. In this case, the earworm was BP's "failed negative test" explanation of the *Deepwater Horizon* blowout.

By repeating the story, and having supposedly credible sources endorse it, BP effectively marketed it: The company promoted and "sold" a false statement to the American media, citizens, and government.

Ignorance

My attorney team—Shaun Clarke, David Gerger, Dane Ball, and David Isaak—was composed of brilliant people, and not one of them had a clue what a negative test on an offshore drilling rig entailed when we began our legal journey together. The same lack of knowledge affected the attorneys for the Department of Justice, and all media, with the notable exception of perhaps one or two from oil industry publications. Even most "experts" assigned to investigative teams that produced the various reports weren't sure what a negative test was, or if they knew what it was, they had likely never conducted one. This was the case with one of the government witnesses at the Federal Oil Spill trial. Even though Richard Heenan, a mechanical engineer, gave testimony on the negative testing done on *Deepwater Horizon*, his published credentials indicated he had zero experience with it on a deepwater drilling rig.

With all these well-educated people in the dark about what well site supervisors did on the rig on April 20, 2010, how could the rest of the country be expected to understand what we were being accused of doing or not doing?

Ironically, the real cause of the blowout is a lot easier to understand. It's a little like the O-ring failure that caused the Challenger Space Shuttle explosion: A piece of equipment didn't do its job.

Politics

In America, citizens can be scapegoated by government prosecutors in response to political pressures. It has happened many times, and no one should be naïve about this reality.

When there is too much centralized control of the Department of Justice and US attorneys, it is highly likely we see the kind of scapegoating that occurred in the BP case. There are those who will defend this kind of control—expressed as some level of intervention and influence from the White House and/or prevailing political party—by saying that it helps create a consistency in policy and in interpretations of law. On the other hand, it can undermine the rule of law.

It's important to note that "influence" can be in the mind of the Attorney General more than it is in the intent of the White House, however, evidence suggests that was not the case during the Obama Administration. There was a coordinated effort between DOJ and the White House to support certain policies and bury others. Protection of the environment was on the winning side.

The Department of Justice that indicted me was Eric Holder's DOJ and it suffered from damaging centralized control—damaging to American justice, that is. Eric H. Holder, Jr. is the first sitting cabinet member that Congress slapped with a contempt citation. His six-year tenure was defined by the politicization of DOJ actions. That's a polite way of saying the DOJ suffered from corruption. His supporters applauded him for doing it, arguing that Holder "used the Justice Department to advance minority rights," among other things.[2] That's nice, except when the DOJ is "used" to further a political agenda, it is abused— and so are the American people.

DOJ's press release about BP's plea agreement captures how misguided the process was. What it says about Don Vidrine and me was false and demeaning. More importantly, the BP guilty plea agreement implicates Don and me by presuming negligence and connecting that presumption with alleged culpability.

In agreeing to plead guilty, BP has admitted that the two highest-ranking BP supervisors onboard the Deepwater Horizon, known as BP's "Well Site Leaders" or "company men," negligently caused the deaths of 11 men and the resulting oil spill. The information details that, on the evening of April 20, the two supervisors, Kaluza and Vidrine, observed clear indications that the Macondo well was not secure and that oil and gas were flowing into the well. Despite this, BP's well site leaders chose not to take obvious and appropriate steps to prevent the blowout. As a result of their conduct, control of the Macondo well was lost, resulting in catastrophe.[3]

After the *Deepwater Horizon* tragedy, the DOJ demanded that BP provide or sacrifice employees for prosecution and BP acquiesced to government demands, breaking its own avowed ethical, moral, and legal standards to preserve long term profitability.

Consider BP's statements in its Sustainability Review 2009, which articulates these commitments and indicates they should have been enforced prior to the blowout:

BP's code of conduct and values demonstrate our commitment to integrity, ethical values and legal compliance.[4]

. . . BP's reputation, and therefore its future, depends on every BP employee, everywhere, every day, taking personal responsibility for ethical and compliant business conduct. It is a fundamental BP commitment to comply with all applicable legal requirements and adhere to high ethical standards.[5]

. . . We have an annual compliance certification process in which all senior level leaders are asked to submit a certificate stating that they personally understand and adhere to the code of conduct and have discussed the code and OpenTalk with their teams. Leaders are also required to report any breaches of the code that occurred in their teams. This process rolls up the management line to the group chief executive, who signs a certificate for the whole group and reports to the board's safety, ethics and environment assurance committee.[6]

It took *two* unethical entities to legally scapegoat innocent BP employees. BP lacked ethics consistent with its own policies and acted with malice. DOJ attorneys accepted the employee sacrifices and assigned legal significance to the false and misleading statements they were told by BP lawyers.

Choosing Scapegoats

What BP did in 2012 was a blueprint—flawed and smeared—of what Volkswagen America tried to do in 2015. A quick look at how Volkswagen CEO Michael Horn handled the company's emissions scandal points to how BP chose its scapegoats three years before.

The year 2015 spelled chaos for Volkswagen – after a scandal involving the deception of customers and emissions regulators it was hit with one of the costliest automotive recalls in history. And to get out of being blamed, CEO Michael Horn played the "rogue employee" card to explain how the cars' engine software cheated in pollution tests.

After talking to the US House Committee on Energy and Commerce, Horn said he only learned of the existence of the "defeat device" when regulators uncovered the software. Much like the rest of the company's senior executives, he claimed management had been in the dark about more than half a million cars having been rigged to suppress emissions while being tested.

"This was not a corporate decision from my point of view," Horn said. "To my knowledge this was a couple of software engineers."[7]

Horn may have had a copy of the BP playbook, but so did the United States Department of Justice—the post-*Deepwater Horizon* new-and-improved version. This time, instead of going to the top and allowing executives to look down the chain of command to finger a scapegoat, the DOJ attorneys gave relatively low-level engineers the chance to provide evidence that it was the executives who were culpable. As a result, Federal prosecutors indicted six executives at Volkswagen, one of whom happened to be in the United States vacationing at the time. Oliver Schmidt, a former emissions compliance executive for Volkswagen Group, was apprehended and on December 6, 2017, he was sentenced to seven years in prison.

Inevitability of Indictments

In June 2011, University of Michigan Law Professor David Uhlmann, a former DOJ lawyer who served as Chief of the Environmental Crimes Section for seventeen years, wrote a faculty paper titled "After the Spill is Gone: The Gulf of Mexico, Environmental Crime, and Criminal Law." In the paper, he made prophetic statements one year before the BP guilty plea agreement and one year before the indictments were announced:

The Justice Department is likely to prosecute BP, Transocean, and Halliburton for criminal violations of the Clean Water Act and the Migratory Bird Treaty Act, which will result in the largest fines

ever imposed in the United States for any form of
corporate crime.[8]

Criminal prosecution of the Gulf oil spill became
inevitable when BP could not stop the flow of oil
from the Macondo well and could not prevent the
resulting oil slick from reaching the shores of the
Gulf of Mexico. Because of the notoriety of the
case, the Gulf oil spill will be seen by many as the
paradigmatic environmental crime.[9]

Uhlmann clearly understood the die had been cast.

Cooperation Credits

Probably about the same time Uhlmann was
committing his prophecy to paper, the DOJ was engaged
in following its common policies and procedures of
offering "cooperation credits" to the corporations involved
in the Macondo disaster to assist the DOJ in identifying
culpable individuals. The invitation to negotiate a deal
trading cooperation credits for culpable BP employees
had been extended. DOJ's probable message to BP
executives via their attorneys: Turn someone in at BP; save
yourself.

The stripped-down concept of cooperation credits is a
basic *quid pro quo*: I do you a favor if you do something
for me. However, the specifics of getting cooperation
credit, according to the Justice Department's own FAQ

on the subject, include a caveat about pointing the finger at real people:

> Under the Policy, a company must turn over all non-privileged relevant information about the individuals involved in the misconduct in order to receive any consideration for cooperation. This is a threshold requirement, and unless it is satisfied, the company will be ineligible for cooperation credit.[10]

This language makes the clear statement that part of the deal is turning in the people you finger as "bad guys." And in case the company evaluating its options didn't get the message the first time, the DOJ FAQ later restates the requirement about "individual involved in the misconduct", and then follows it with a note on how to get the most ideal outcome:

> The actual cooperation credit that a company ultimately receives, however, will depend on a number of additional factors. These include the timeliness of the cooperation, the diligence, thoroughness and speed of the internal investigation, and the proactive nature of the cooperation.[11]

Complicity of the Board

People who served on the BP Board of Directors during or shortly after the tragedy were briefed throughout the

BP-DOJ settlement negotiation process and approved the deal. The signatures of company officers on the BP Guilty Plea Agreement affirmed that board members in November 2012 unequivocally knew the terms. Exhibit C entitled "Certification of Resolutions Adopted by the Board of Directors of BP Exploration & Production Inc" is an admission that those individuals were willing to put at risk BP employees who could have languished in prison for many years. All the Resolutions start with the usual "Whereas…" but only one makes it perfectly clear that BP's board members were complicit in the scapegoating:

WHEREAS, the Board acknowledges that the Plea Agreement fully sets forth the Company's agreement with the United States with respect to all criminal violations identified during the Investigations and that no additional promises or representations have been made to the Company by any officials of the United States or the States in connection with the disposition of the Investigations, other than those set forth in the Plea Agreement.[12]

What followed is the commitment that sealed the deal.

RESOLVED that:
1. The Board approves and agrees that it is in the best interest of the Company to enter the guilty plea provided for, and agrees to the terms

provided in the Plea Agreement with the United States Department of Justice in substantially the form and substance set forth in the form of Plea Agreement presented to this Board.[13]

The board members who acceded to this, and are still on the Board of Directors, are:
Carl-Henric Svanberg, Chairman of Board since January 1, 2010
Robert Dudley, Group Chief Executive (British equivalent of CEO) since October 1, 2010
Dr. Brian Gilvary, Chief Financial Officer since January 2012
Paul Anderson, who joined the BP board 2010
Admiral Frank Bowman, who served in the US Navy for 38 years and joined the BP board November 2010
Ian Davis, who joined the board April 2010
Brenden Nelson, who joined the board November 2010
Professor Dame Ann Dowling, who joined the board February 2012

I mention when they joined the board to spotlight the fact that none was new to his or her director responsibilities or to BP policies at the time they endorsed the guilty plea agreement, which formed the basis for the indictments against two BP employees, Don Vidrine and me.

The following are no longer on the BP board, but were at the time of the guilty plea agreement:

Anthony Burgmans
Cynthia Carroll
Iain Conn
George David
Bryon Grote
Phuthuma Nhleko
Andrew Shilston

The above are the people entrusted to establish the company's ethical standards, and presumably to uphold them as well. They did not.

Complicity of BP Executives/Managers

Numerous BP executives gave false statements from former BP CEO Tony Hayward on down. Several even lied under oath in hearings and in courtrooms and are referenced in previous chapters. Here, I just want to single out two corporate people who built the case for the indictments against the well site leaders.

Both of these men were leaders of the BP Accident Investigation Team and approved the publication of several key inaccuracies in the final BP Accident Investigation Report. This report influenced other investigation teams as well as media coverage in the aftermath of the tragedy. It was the first official investigative report and people looked to it for explanations. In addition, media got caught-up in their key assertions. I want to re-emphasize that even people on the other investigation teams were not necessarily experts in deepwater offshore drilling, so they were relying heavily on BP's investigation team insights. For those other teams, it was not a matter of deliberately propagating lies, but one of not being able to distinguish fact from fabrication.

The two people were, at the time, the Group Head of Safety & Operations, and Vice President, BP Exploration (Alaska).

The former, Mark Bly, was appointed to lead the BP Accident Investigation Team. He allowed several misleading or false statements to be published in the BP Accident Investigation Report. Unfortunately, those erroneous statements changed the content and direction of the entire Macondo investigation process. Eventually, Bly got caught-up in one of the false statements he allowed to be published in the BP Accident Investigation Report— and it was on the witness stand during the Federal Oil Spill trial.

The latter, Steve Robinson, became a leader of the BP Accident Investigation Team. He, too, allowed several false statements to be published in the BP Accident Investigation Report and they changed the content and direction of the Macondo investigation process. And as was the case with Bly, he got caught during the Federal Oil Spill trial.

Nothing but the Truth

The single direct cause of the Macondo blowout was because the blowout preventers failed to seal and contain the Macondo well kick. BP has never admitted that.

After the DOJ went 0-5 for felony convictions against Macondo defendants—technically 0-48 if all of the individual indictments against the four BP employees are considered—no one in America, not the media, nor the Legislative or the Executive branches of the United States Government demanded that the DOJ examine how and why the DOJ failed completely in its *Deepwater Horizon* prosecutions. No one demanded that the government apologize to two citizens who were scapegoated to meet political and corporate needs. So there was no examination and there was no apology.

The US DOJ strategy of offering cooperation credit did not yield the truth and it did not yield justice.

In the end, it is easy for me to state that politics superseded American justice; I was aware of it and had to endure it for many years. Here is the crucial fact: If you are the one unjustly accused, Justice—capital J—is never served.

TECHNICAL SPECIFICATIONS OF THE BOP AND REASONS FOR FAILURE

The purpose of this appendix is to provide more technical detail about the equipment that was out of Code of Federal Regulation compliance and failed on April 20. It is not story-focused, although it is impossible to leave the narrative behind completely.

The Blowout Prevention Equipment on Deepwater Horizon on April 20, 2010

The BOP equipment (BOPs) was installed on *Deepwater Horizon* and certified by Cameron International Company on December 13, 2000. The stack was 53-feet tall, weighed approximately 700,000 pounds, had a rated working pressure of 15,000 psi and consisted of, from top to bottom, two annular preventers, a Lower Marine Riser Package (LMRP) connector, one blind shear ram (BSR), one casing shear ram (CSR), two variable sized pipe rams (VBR) (3 ½ inches to 6 5/8 inches range) installed to hold pressure from below the ram only, and one variable pipe ram (VBR) flipped over to hold pressure from above the ram. The top of the BOPs at the riser adapter was at an ocean depth of 5,001feet and the bottom of the BOPs latched into wellhead at the ocean floor mudline was at an ocean depth of 5,054 feet measured from the rig floor.

The stack consisted of three types of independent and redundant closing devices: two annular preventers, two types of shear rams and three variable bore pipe rams. On the evening of the blowout, April 20, 2010, five of the seven closing rams and elements could have sealed the Macondo well kick if those closing rams and elements had functioned properly as designed.

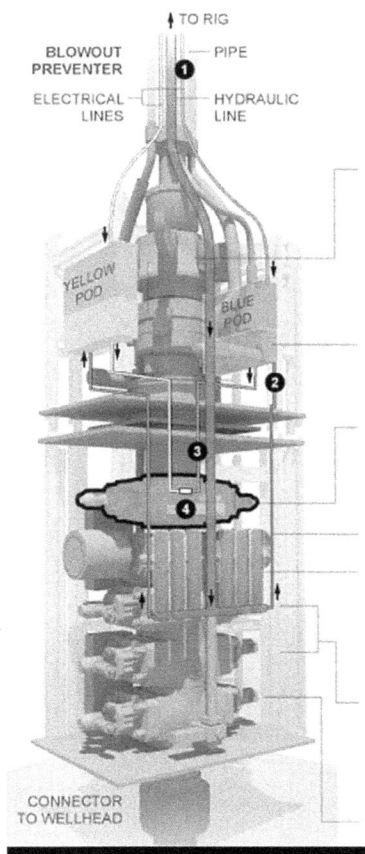

The BOP stack is 53 feet tall.

Annular Preventers
Designed to create a seal around the drill pipe or seal off an open wellbore when there is no pipe.

Control Pods
Receive electrical signals from the rig and direct the movement of hydraulic fluid. Upper portion has electrical parts; lower portion has hydraulic valves. Only one pod is activated at a time.

Blind Shear Ram
Cuts drill pipe and completely seals the well.

Casing Shear Ram
Cuts drill pipe in an emergency when the rig must disconnect from the well quickly.

Accumulators
Store fluid sent from the rig. During an emergency, pressurized fluid from these canisters can provide force to power the blind shear ram.

Pipe Rams
Seal off space between outside of the drill pipe and well bore to keep pipe centered.

Test Ram
Used to test the rams above it.

1. The line on the right indicates AC electrical line from the drilling rig to the Control Pod

2. The thicker line in the center is a hydraulic pressure sensor from the subsea BOP accumulator system to the Control Pod

3. The thin line just to the left of 2. Is a command signal from the blue pod to open or close the Blind Shear Ram (BSR)

Important: There are redundant signal lines from the blue pod and the yellow pod to the BSR (see 4). The redundant signal lines are important to understand how the fail-safe AMF/deadman system works. When the electrical signal as well as hydraulic pressure and fiber-optic communication from the drilling rig (see 1) is lost completely the blue and yellow pods will automatically activate from battery power the AMF/deadman system, which will close the BSR. The AMF/deadman failed on the evening of April 20, 2010 because the 27 V battery in the blue pod was depleted and an electrical signaling solenoid valve in the yellow pod was mis-wired.

4. (yellow and blue signal lines terminating at the Blind Shear ram) reveals the redundancy of the Yellow and Blue Control Pods. When one pod fails the redundant back-up pod will activate the same function.

What are annular preventers? And why are annular preventers different from blind and casing shear rams and variable bore ram preventers?

The most obvious difference between an annular preventer and the other types of preventers is that an annular preventer has a closing element inside the high-pressure housing consisting of a flexible rubber element while the other types are primarily solid steel bodied closing rams.

An annular preventer is the most multi-purpose preventer of the three types. All three types of closing devices are enclosed in high pressure housings (rated to 15,000 psi on the Deepwater Horizon BOPs) The housing of an annular preventer is shaped much like a upside down light bulb where the top of the housing is rounded so the top of the mostly rubber closing element with steel ribbing of the annular rubber element will follow the contour of the preventer housing when closed. Inside the housing is a fairly simple closing/opening system. There is either a flat or wedge type piston in the lower section of the housing. The donut shaped rubber element sits on top of the piston.

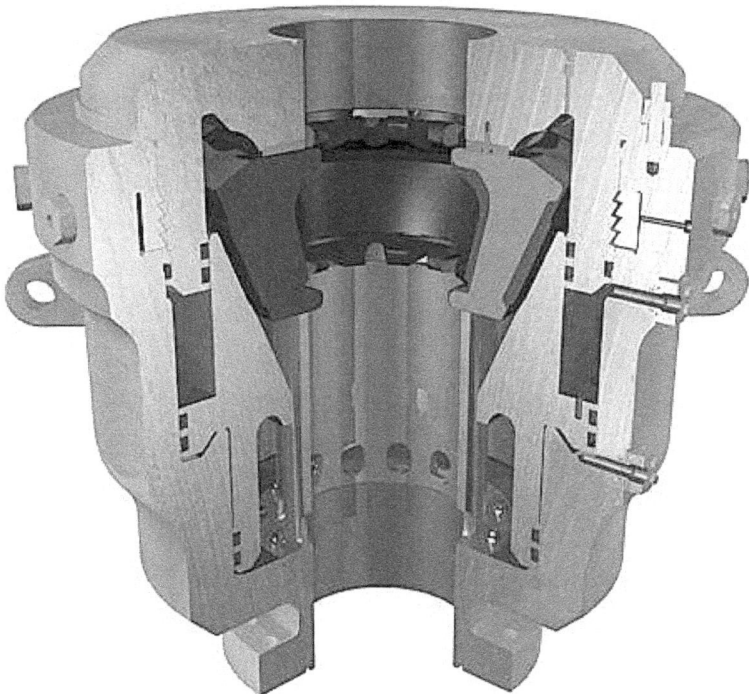

Hydraulic fluid under pressure pumped through a closing port below the piston, which causes the piston to push up, squeezing and deforming the rubber element into the wellbore area of the BOP. To open the annular preventer, hydraulic fluid under pressure is pumped into an opening port that causes the piston to move down thereby allowing the rubber element to return to its original donut shape.

Annular rubber elements can deform, squeeze, and seal against almost anything or nothing in the BOP stack including various sizes of casing, drill pipe, drill pipe collars, odd shaped tubular components or wire line in the well bore area of the BOP. Additionally, they can squeeze into the wellbore area with nothing in the well and still seal a well kick. The driller on duty is responsible to shut-in a well if he detects a kick. If he is uncertain about the location of a large drill collar in the BOP or if he is shutting in a well from a remote closing station, he will likely chose to close an annular preventer versus a VBR because a squeezed annular rubber will not damage any tubular in the BOPs.

Because the rubber packing element is flexible, the annular preventer rubber can also be used for *stripping*. Occasionally drill pipe must be run in or out of the wellbore while holding back well pressure using a closed preventer. Closing pressure on the annular rubber element can be reduced and the drill pipe can be raised or lowered past the rubber element while still holding back well pressure. Stripping pipe through an annular rubber can, and often does, damage and prematurely wear-out packing elements. Approximately six weeks before the upper annular rubber was activated to seal the Macondo kick, the rig crew used the upper annular rubber for stripping. The rig crew stripped approximately 2,209 feet of drill pipe consisting of 48 tool joints through the closed upper annular rubber which did cause damage to the upper annular rubber. Rig crew members testified at hearings that they observed pieces of annular rubber come out of the well during circulation of the well after the stripping operation.

Why the upper annular preventer was activated for the blowout and not the lower is not clear. After the March 8 "well control event" on Deepwater Horizon, Halliburton's OpenWells—Operations Reporting recorded: "Stripped drill pipe through upper annular preventer from 17,146 ft. to 14,937 ft. while addressing wellbore losses."[1]

With a standard drill string length of 46 feet per section, approximately 48 tool joints were stripped through the upper annular rubber while it was closed. As SINTEF, the largest independent research organization in Scandinavia, observed in a previous study, ". . . experience shows that when stripping is required as a part of the kick killing operation, this will cause that the annular is likely to fail afterwards."[2] Despite the annulars having received hard use and being in need of maintenance, the official word from BP was "we don't want to the change the annulars. . ." The official corporate response was "BP accepts responsibility if both annulars were to fail."[3]

Casing Shear Rams and Blind Shear Rams

The *Deepwater Horizon* BOP stack contained two types of shear rams: a blind shear ram (BSR) and a casing shear ram. Shear rams are the least used type of BOP preventer rams because they are generally used for positive casing pressure tests and activated in emergencies only.

BSRs are designed to cut or shear the body of drill pipe in the BOP first, and then to seal the well after the drill pipe is cut. Blind shear rams are designed to seal pressures as high as Maximum Anticipated Wellhead Pressure (MAWHP), that is, the highest pressure expected to be encountered at the wellhead. The BSR in the Deepwater Horizon BOP was never pressure tested to MAWHP, however. For that reason, it was unknown if the BSR could have sealed a Macondo well kick.

On *Deepwater Horizon*, the single BSR was used non-emergency functions prior to the blowout. Rig crews used the BSR to conduct positive casing tests where the BSR was closed and drilling fluid pumped below the ram; that, in turn, caused casing pressure in the casing to increase below the ram. The BSR was never regularly pressure tested to the same pressure standard and or regularity as the Variable Bore Rams. The BSR was once pressure tested to only 914 psi while on the Macondo well. And the BSR held 2,600 psi during the positive casing test on April 20, 2010.

It's also important to note there have been conflicting opinions in investigative reports as to whether or not the BSR was capable of cutting the body of 6 5/8-inch, 32.7 pounds-per-foot drill pipe that was used while drilling the Macondo well.

Casing shear rams (CSRs) are designed to shear casing and other large tubular components in a BOP stack, but not designed to seal a well after cutting the casing. The CSR was not activated to control the Macondo well kick.

There were five different methods to activate the BSR on the rig on April 20, 2010. Evidence shows that the BSR was activated by two separate activation methods on April 20; the Emergency Disconnect System (EDS) was activated at 9:56 pm and the AMF/deadman was automatically activated at 9:49 pm. The BSR did not cut the 5 ½", S-135 grade drill pipe and did not seal the Macondo well kick as designed. It is unknown why the BSR did not cut and seal the Macondo well. We know the BSR was never pressure tested to Maximum Anticipated Wellhead Pressure (MAWHP) while on the Macondo well. Because the blind shear ram was never pressure tested to MAWHP no one can ever know if the blind shear rams could have sealed the Macondo kick.

Variable Bore Rams (VBR)

VBR are solid steel bodied rams designed to close around drill pipe, some types of heavy weight drill pipe and sometimes around small drill collars and seal all high pressure well kicks. The reason the rams are called *variable* bore rams is they are designed to close around a range of sized drill pipe and not just one specific sized pipe. As an example, one VBR range may be to close around 3 ½-inch, 4-inch, 4 ½-inch, 5-inch and 6 5/8-inch drill pipe. The VBR is designed with overlapping concentric fingers similar to an airport baggage conveyor that allows the ram to close around variable drill pipe sizes and seal a well kick.

Variable bore rams are generally designed to hold higher well kick pressures than annular rubbers and therefore are regularly pressure tested to the highest BOP test pressures. Usually a VBR is designed to hold as much pressure as the rated working pressure of the BOP stack. At Macondo, the blind shear ram, and the upper and middle variable rams were manufactured and designed to hold 15,000 psi of well pressure. The lowest VBR was flipped over to use as a BOP test ram. All preventer rams and rubbers are designed to hold pressure from only one direction, therefore the bottom VBR had to be flipped over in order to conduct pressure testing inside of the BOP. All BOP rams and rubbers are not designed to hold pressure from above the ram or rubber. This design fact was important in the discussion of the negative tests.

Both the upper and middle variable bore rams were activated on April 20, 2010 at 9:47 p.m. and both variable bore rams failed to seal the Macondo well kick. The upper and middle VBR rams were original BOP equipment certified on December 13, 2000 and had never been completely disassembled and inspected to meet federal regulatory regulations prior to the Macondo blowout. Additionally, the VBRs were not regularly pressure tested to MAWHP while drilling the lowest formations of the Macondo well.

Blowout Prevention Equipment is designed, manufactured and expected to be "fail-safe" on offshore deepwater drilling rigs.

The subsea BOP Automatic Mode Function ("AMF") or "deadman" system is designed to activate and close the Blind Shear Ram when all three of the following conditions are met:

- the Lower Marine Riser Package (LMRP) suffers the loss of electrical power
- the loss of fiber-optic communication with the rig
- and the loss of hydraulic pressure from the rig.

The accumulator bottles located on the lower BOP stack provide the hydraulic power for the AMF (see illustration above).

The AMF relied upon two redundant control pods—a blue pod and a yellow pod. Under normal operations, the pods are powered through AC cables from the surface. In the event of a loss of power from the surface, the power supply for each of the control pods is maintained through batteries located in the subsea electronics module (SEM) in the multiplex section of each pod. The pods were located on opposite sides of the LMRP. Each function independently and each has its own power supply and batteries. Each pod includes solenoid valves, which are devices that open and close in response to electrical signals. The solenoids are designed to communicate with the BOP elements and trigger the delivery of 4,000 psi closing pressure to the BSRs through the dedicated accumulator bottles located on the lower BOP stack.

Based on Macondo investigation evidence, it is likely that the explosion on April 20 created loss of electrical power, loss of fiber-optic communication and loss of hydraulic pressure. The AMF/deadman likely activated at 9:49 pm on April 20 but failed due to improper inspection and maintenance of the *Deepwater Horizon* BOPs.

The AMF/deadman failed on April 20 because of a dead 27 volt battery in the blue control pod and a mis-wired Solenoid 103 Y in the yellow control pod. Both failures were due directly to deliberate improper maintenance.

**

I have heard the question several times: Were BP's negligent actions to save money typical in the industry? Obviously, there is no way for me to know that for sure, but two points help suggest an answer.

First, the importance of maintaining blowout prevention equipment has been engrained in industry practices for decades. Blowout preventers have been standard safety equipment in the drilling industry since shortly after they were introduced by Cameron, the manufacturer, in 1922. Cameron's invention is credited with revolutionizing operations in an industry that was in its infancy in the early part of the twentieth century. The American Petroleum Institute (API)—the leading force behind oil and gas industry standards for more than ninety years—published its first set of recommended practices for maintaining blowout prevention equipment systems for drilling wells in February 1976. As is common for the industry, API's work later became the Code of Federal Regulations. In other words, the value and necessity of maintaining blowout preventers have been recognized by the industry for decades and the requirement to do so by the federal government followed.

Second, we can look to the recorded instances of successful prevention of blowouts as indicators that compliance is more common than non-compliance. According to David Uhlmann, the former chief of the Justice Department's environmental crimes section:

In the United States, there were twenty-eight major drilling related spills, natural-gas releases, or well-control incidents in the Gulf of Mexico during 2009, including a loss of well control and an explosion in April that did not result in a major spill only because the blowout preventer worked. No criminal charges were filed.[4]

GLOSSARY

Annular blowout preventer – The primary blowout prevention mechanism featuring a kind of deformable rubber doughnut that squeezes tight until it creates a seal.

Automatic Mode Function (AMF)/deadman – A blowout prevention closing system that comes into play when all electric power and pressure communication is lost from the rig to the subsea Blowout Preventer (*see* Blowout Preventer); it is the final blowout prevention system that automatically activates the blind shear ram to cut through the pipe and seal the well kick so oil/gas stop moving toward the rig.

Blind Shear Ram (BSR) – A high pressure blowout prevention mechanism designed to both cut pipe in the BOP and simultaneously seal the well kick when the blind (flat faced) rams close.

Blowout Preventer (BOP) – A massive high-pressure rated mechanism designed to stop unexpected fluid flow (*see* Kick) from the well to the rig; a BOP is also designed to regain control of a well kick.

Driller – A drilling contractor team leader in charge of the process of well drilling.

Drilling and Well Operations Practice (DWOP) – The manual issued to selected BP employees to guide the users in enforcing BP minimum drilling standard procedures and reducing significant risks

Drilling fluids specialist – (*See also*, Mud engineer) a person responsible for testing the mud at a rig and for prescribing mud treatments to maintain proper mud weight, properties and chemistry within recommended parameters.

Emergency Disconnect System – the system that allow a deepwater offshore drilling rig and marine drilling riser to disconnect at the BOP, thereby allowing the rig to become a sailing vessel

Involuntary manslaughter – Generally speaking, the legal term to describe unintentional killing resulting from recklessness or criminal negligence, or from some unlawful act such as driving while intoxicated.

Kick – The relatively common event that occurs when the drill bit enters what is called an "over-pressured" formation; that formation overpressure will exceed the weight of the fluid column in the well and begin to push fluid up the well.

Mud – Drilling fluid used in hydrocarbon drilling operations; can be water-based, oil-based or synthetic-based drilling fluids (*see also,* Synthetic Oil Based Mud)

Mud engineer – (*See also,* Drilling fluids specialists) a person responsible for testing the mud at a rig and for prescribing mud treatments to maintain proper mud weight, properties and chemistry within recommended parameters.

Negative Test – A test to confirm well integrity.

Offshore Installation Manager (OIM) – The top drilling official on an offshore drilling rig; the OIM can override anyone's decision on the rig.

Positive Test – A test to confirm well integrity.

Rig auditor – A member of a rig audit team employed to conduct a comprehensive audit of drilling rig equipment and compliance with safety management systems often authorized by international oversight organizations to annually conduct technical inspection and evaluation of registered merchant vessels and offshore drilling rigs.

Seaman's Manslaughter –A statute of the US Criminal Code (918 U.S.C. § 1115) that criminalizes misconduct or negligence of people who have a professional role onboard a seagoing vessel.

Synthetic Oil Based Mud (SOBM) –A specially formulated drilling fluid.

Toolpusher – A drilling contractor rig supervisor in charge of drilling contract rig personnel and process of rig operations.

Variable Bore Ram (VBR) – A high-pressure, steel bodied, ram-type blowout prevention mechanism; "variable" means that the ram can close and seal around a range of different pipe sizes.

Well site leader – A professional designation by BP for someone assigned onsite a drilling rig who has job roles to safely and effectively implement the approved drilling well plan in coordination with all contracted personnel and equipment.

ENDNOTES

Introduction: Fact and Fiction

1. Matt Brennan, *LA Weekly* film critic, in an interview with WWL TV reporter David Hammer for the story "Deepwater Horizon movie: separating fact from fiction", September 21, 2016; http://www.wwltv. com/entertainment/deepwater-horizon-movie-separating-fact-from-fiction/322172349

2. Maryann Karinch, *Nothing But the Truth* (Wayne, NJ: Career Press, 2015), p. 27.

3. Josh Rottenberg, "Telling the true story of the heroes behind 'Deepwater Horizon,' without using 'stolen valor'," *Los Angeles Times*, September 22, 2016; http://www.latimes.com/entertainment/movies/la-ca-mn-deepwater-horizon-feature-20160914-snap-story.html

Chapter One

1. Jesse Eisinger, *The Chickenshit Club: Why the Justice Department Fails to Prosecute Executives* (New York: Simon & Schuster, 2017)

2. http://fuelfix.com/blog/2015/07/20/the-number-of-temporarily-sealed-wells-in-the-gulf-of-mexico-is-growing/

3. http://www.nola.com/news/gulf-oil-spill/index.ssf/2010/07/27000_abandoned_oil_and_gas_we.html

4. API RP 53 18.10.3

5. From an email exchange between John Guide, BP, and Paul Johnson@deepwater.com on April 15, 2010.

6. *Ibid*

7. David Whitford, Doris Burke, Peter Elkind, BP: 'An accident waiting to happen', Fortunte, January 24, 2011; http://fortune.com/2011/01/24/bp-an-accident-waiting-to-happen/

8. *Ibid*

9. Request email sent by Powell, Heather (JC Connor Consulting), Fri Apr 16 15:25:58 2010 to 'frank.patton@mms.gov', Subject MC252#1 TA APM; Approval sent by frank.patton@mms.gov, Fri Apr 16 15:47:27

2010 to Powell, Heather (JC Connor Consulting), Subject Permit to Modify Well at Surface Location Lease: G32306 Area : MC Block: 252 Well Name: 001 with API Number: 608174116901 has been approved

10. David Hammer, "Updates from oil rig explosion hearings: MMS engineer admits he approved blowout preventer without assurances it would work." The Times-P_icayune, May 11, 2010

11. Earl E. Devaney, Inspector General, United States Department of the Interior, in a memorandum to Secretary Dirk Kempthorne, dated September 9, 2008 and released to the public September 10, 2008, page 2

12. *Ibid*

13. *Ibid*

14. Drilling and Well Operations Practice, Rev 6 (October 2008), BP, Part A, 1.4, p. 12

Chapter Two

1. Schlumberger Oilfield Glossary; http://www.glossary.oilfield.slb.com/en/Terms/r/roustabout.aspx
2. *Ibid*; http://www.glossary.oilfield.slb.com/Terms/r/roughneck.aspx
3. "Short portraits of 11 who died on the Deepwater Horizon," *Clarion Ledger*, April 18; 2015http://www.clarionledger.com/story/news/2015/04/18/short-portraits-died-deepwater-horizon/26007421/
4. *Ibid*
5. Schlumberger Oilfield Glossary; http://www.glossary.oilfield.slb.com/Terms/m/mud_engineer.aspx
6. *Ibid*; http://www.glossary.oilfield.slb.com/Terms/t/toolpusher.aspx
7. "Short Portraits", *Ibid*

Chapter Three

1. Attached to the top of a string of casing that is run into an oil and gas well is a precisely sized portion of the wellhead assembly called a casing hanger. The casing hanger is an exact fit and lands on a shoulder inside the wellhead. Casing hangers are designed to hold the entire weight of the casing and provide a seal between the casing hanger and the wellhead. Often in cased oil and gas wells, sealing packers between two different sizes of casing are set near the bottom of the hung-off casing string, so when warm oil is produced through hung-off casing the string can expand and contract due to temperature differences which could push a casing hanger off its wellhead shoulder and seal. It is also possible that gas pressure could push a casing hanger off its wellhead shoulder and seal. Therefore, to ensure that a casing hanger remains set in-place, a lockdown sleeve is set over the hanger. Think of a lockdown sleeve like the wire cage over a champagne bottle. The wire

cage prevents the champagne cork from popping out of the bottle if the bottle is shaken and the gas inside is excited.

2.   Fred H. Bartlit, Jr., Chief Counsel; Sambhav N. Sankar, Deputy Chief Counsel; and Sean C. Grimsley, Deputy Chief Counsel; "Macondo: The Gulf Oil Disaster", Chief Counsel's Report, 2011, National Commission on the BP Deepwater Horizon Oil Spill and Offshore Drilling; http://www.wellintegrity.net/documents/ccr_macondo_disaster.pdf

3.   *BP Accident Investigation Report*, Appendix Q page 2.

4.   The first negative flow test via the drill pipe was conducted per common practice on the *Deepwater Horizon* and using the exact procedure used on the Tiber Well (August 25, 2009) and on the Kodiak Well January 28, 2010). On the Tiber Well, drill pipe depth was 5,864 feet, final circulating shut-in pressure was 1,690 psi, and bleed-off volume was 9.5 barrels. On the Kodiak Well, drill pipe depth was 5,470 feet, final circulating shut-in pressure was 1,820 psi, and bleed-off volume was 11 barrels.

5.   Three barrels of seawater flowed out of the drill pipe during the flow check likely due to kill line overbalance.

6.   While researching all negative tests ever conducted on the *Deepwater Horizon* I discovered there were two negative tests that were observed for flow for five minutes and then deemed successful. Five minute negative tests had been conducted on the *Deepwater Horizon* in the past.

7.   In his investigative report for *Popular Mechanics*, Carl Hoffman also confirmed that Jimmy Harrel deemed the tests to be successful. Carl Hoffman, "Special Report: Why the BP Oil Rig Blowout Happened," Popular Mechanics, September 2, 2010; https://www.popularmechanics.com/science/energy/a6065/how-the-bp-oil-rig-blowout-happened/

8.   Ibid, Chief Counsel Report, p 242

Chapter Four

1.   Walter Pavlo, "Two Years After Ruling, BP Engineer Still Carries Burden of Prosecution," Forbes, January 2, 2018; https://www.forbes.com/sites/walterpavlo/2018/01/08/two-years-after-ruling-bp-engineer-still-carries-burden-of-prosecution/

2.   David Barstow, David Rohde and Stephanie Saul, "Deepwater Horizon's Final Hours," The New York Times, December 25, 2010; http://www.nytimes.com/2010/12/26/us/26spill.html?pagewanted=all

3.   "Interview of Bob Kaluza on Board the *Damon Bankston*," April 21, 2010, Questioners: Investigator Glynn Breaux, U.S. Department of Interior, MMS; Lieutenant Angelique Flood, U.S. Coast Guard; and Specialist Randy Josey, U.S. Department of Interior, MMS; Present, asked no questions: Cliffe Laborde, Attorney for Tidewater, Inc.; and Lieutenant Barbara Wilk, U.S. Coast Guard, pp 15-16 of 47.

4.   *Ibid*, p 7

Chapter Four

1.   https://www.linkedin.com/in/patobryan1/
2.   GoM D&C Development Well Delivery RACI Chart. Rev 12, October 6, 2009. R=Responsible, A=Accountable, C=Consulted, I=Informed
3.   From: Guide, John; Sent: Wednesday, October 07, 2009 6:40 AM; To: Cocales, Brett W; Daigle, Keith G
4.   Subject: FW: Deepwater Horizon Rig Audit From Defendant's Exhibit 95 in the February 2016 trial of Robert Kaluza
5.   Testimony of Harry Thierens to U.S. Coast Guard & Bureau of Ocean Management Joint Investigation – Houston, August 25, 2010; https://www.c-span.org/video/?295172-2/deepwater-horizon-joint-investigation-harry-thierens-testimony-part-1
6.   From an email exchange between John Guide, BP, and Paul Johnson@deepwater.com on April 15, 2010.
7.   United States Coast Guard Report of Investigation into the Circumstances Surrounding the Explosion, Fire, Sinking and Loss of Eleven Crew Members Aboard the MOBILE OFFSHORE DRILLING UNIT DEEPWATER HORIZON In the GULF OF MEXICO April 20 – 22, 2010, page 90
8.   *Ibid*, p 95

Chapter Five

1.   Judith Sherwin, Attorney at Law, Adjunct Professor, Loyola School of Law, as quoted by Jade Greear in "Speaking Truth to Power," Huffington Post, December 22, 2015; https://www.huffingtonpost.com/jade-greear/speaking-truth-to-power_2_b_8824094.html
2.   "In Flint Water Crisis, Could Involuntary Manslaughter Charges Actually Lead to Prison Time?" States New Service, June 19, 2017.
3.   Gregory Hartley and Maryann Karinch, "How and Why Do People Lie?", *How to Spot a Liar* (Pompton Plains, NJ: Career Press, 2012)
4.   *Ibid*, p 43
5.   Deepwater Horizon Accident Investigation Report, September 8, 2010, p 89; https://www.bp.com/content/dam/bp/pdf/sustainability/issue-reports/Deepwater_Horizon_Accident_Investigation_Report.pdf
6.   *Ibid*, p 39
7.   "Deep Water: The Gulf Oil Disaster and the Future of Offshore Drilling," Report to the President, National Commission on the BP Deepwater Horizon Oil Spill and Offshore Drilling, January 2011, p 119
8.   Deepwater Horizon Accident Investigation Report, *Ibid*, p 88
9.   *Ibid*, p 40

10. *Ibid*, p 87

11. *Ibid*, Appendix Q, p 8

12. *Ibid*, p 85

13. United States Coast Guard, "Report of Investigation into the Circumstances Surrounding the Explosion, Fire, Sinking and Loss of Eleven Crew Members Aboard the MOBILE OFFSHORE DRILLING UNIT Deepwater Horizon in the Gulf of Mexico April 20-22, 2010," p 96; https://www.bsee.gov/sites/bsee.gov/files/reports/safety/2-deepwaterhorizon-roi-uscg-volume-i-20110707-redacted-final.pdf

14. *Ibid*, p 110

15. Transocean press release, "Transocean Ltd. Announced Release of Internal Investigation Report on Causes of Macondo Well Incident," June 22, 2011 (MARKETWIRE via COMTEX); http://www.deepwater.com/news/detail?ID=1576865

16. The Bureau of Ocean Energy Management, Regulation and Enforcement "Report Regarding the Causes of the April 20, 2010 Macondo Well Blowout," September 14, 201, p 72; https://www.bsee.gov/sites/bsee.gov/files/reports/blowout-prevention/dwhfinaldoi-volumeii.pdf

17. *Ibid*, pp 90-91

18. National Academy of Engineering and National Research Council report of December 14, 2011, p 25

19. The CSB's final report was released on the sixth anniversary of the blowout, April 20, 2016.

20. US Chemical Safety and Hazard Investigation Board, "Investigation Report Overview: Explosion and Fire at the Macondo Well (11 Fatalities, 17 Seriously Injured, and Serious Environmental Damage)", Report No. 2010-10-I-OS, June 5, 2014; http://www.csb.gov/assets/1/7/Overview_-_Final.pdf

21. "Macondo: The Gulf Oil Disaster, Chief Counsel's Report, 2011," National Commission on the BP Deepwater Horizon Oil Spill and Offshore Drilling p I; http://www.wellintegrity.net/documents/ccr_macondo_disaster.pdf

22. *Ibid*, p 218

23. Fred Bartlit, Jr. as recorded by CBS News during the two days of hearings November 8-9, 2010 and reported online in "Gulf Probe: No Sign Cost Was Chosen Over Safety," November 8, 2010; https://www.cbsnews.com/news/gulf-probe-no-sign-cost-was-chosen-over-safety/

24. Chief Counsel Report, *Ibid*, p 246

Chapter Six

1. Andy Rowell, "What the hell did we do to deserve this?" *Oil Change International*, April 30, 2010; http://priceofoil.

org/2010/04/30/%E2%80%9Cwhat-the-hell-did-we-do-to-deserve-this%E2%80%9D/
2. "BP boss admits job on the line over Gulf oil spill," *The Guardian,* May 13, 2010; http://www.bbc.com/news/10360084 https://www.theguardian.com/business/2010/may/13/bp-boss-admits-mistakes-gulf-oil-spill
3. Tony Hayward in an interview with Sky News, as reported by BBC News, "BP boss Tony Hayward's gaffes," BBC News, June 20, 2010;
4. https://www.youtube.com/watch?v=MTdKa9eWNFw
5. "A history of BP's US disasters," Reuters, November 15, 2012
6. *Ibid*
7. *Ibid*
8. BR Staff, "BP's spill probe ignored management failures, safety expert says," *Greater Baton Rouge Business Report,* February 28, 2013; https://www.businessreport.com/article/bps-spill-probe-ignored-management-failures-safety-expert-says
9. Richard Thompson, "BP's safety chief testifies that internal report did not probe management's role in Gulf oil spill," The Times-Picayune, February 28, 2013; http://www.nola.com/news/gulf-oil-spill/index.ssf/2013/02/bps_safety_chief_testifies_tha.html
10. BP Deepwater Horizon Accident Investigation Report, September 8, 2010, p. 13.
11. *Ibid*, unnumbered first page
12. Chief Counsel's Report, *Ibid*, p i
13. "'Everybody hates me' says BP oil spill investigator," The Telegraph, November 8, 2010; http://www.telegraph.co.uk/finance/newsbysector/energy/oilandgas/8118435/Everybody-hates-me-says-BP-oil-spill-investigator.html
14. Judge Carl Barbier, "Findings of Fact and Conclusions of Law: Phase One Trial" In re: Oil Spill by the Oil Rig "Deepwater Horizon" in the Gulf of Mexico, on April 20, 2010, Case 2:10-md-02179-CJB-SS Document 13381-1 Filed 09/09/14, p 71; http://www.uscourts.gov/courts/laed/9092014RevisedFindingsofFactandConclusionsofLaw.pdf
15. *Ibid*, p 72
16. *Ibid*, p 72
17. BP Accident Investigation Report, Ibid, p. 88
18. David M. Uhlmann, "After the Spill Is Gone: The Gulf of Mexico, Environmental Crime, and the Criminal Law," June 2011, para 246; https://deepblue.lib.umich.edu/bitstream/handle/2027.42/107454/109MichLRev.pdf.txt;jsessionid=FEE96701B6A2568C26D93BD8F7D04D87?sequence=3
19. *Ibid*, para 203-206
20. Clifford Krauss, "In BP Indictments, U.S. Shifts to Hold Individuals Accountable," The New York Times, November 15, 2012; http://

www.nytimes.com/2012/11/16/business/energy-environment/in-bp-indictments-us-shifts-to-hold-individuals-accountable.html

21.   Rick Jervis and Kevin Johnson, "3 BP executives indicted over Gulf oil spill," USA Today, November 15, 2012; https://www.usatoday.com/story/money/business/2012/11/15/bp-near-settlement-with-us-over-gulf-spill/1706209/

22.   https://seekingalpha.com/article/962921-bp-management-discusses-q3-2012-results-earnings-call-transcript

23.   Sally Yates, Memorandum of September 9, 2015 from the U.S. Department of Justice, Office of the Deputy Attorney General, p 2; https://www.justice.gov/archives/dag/file/769036/download

24.   United States of America v. BP Exploration & Production, Inc., Guilty Plea Agreement, p 15; https://www.justice.gov/iso/opa/resources/43320121115143613990027.pdf

25.   Clifford Krauss and John Schwartz, "BP Will Please Guilty and Pay Over $4 Billion," The New York Times, November 15, 2012; http://www.nytimes.com/2012/11/16/business/global/16iht-bp16.html

26.   Robert B. Reich, "Why BP Isn't a Criminal," November 16, 2012; http://robertreich.org/post/35848994755;

Chapter Seven

1.   Eric Fielding, Andrew W. Lo, and Jian Helen Yang, "The National Transportation Safety Board: A Model for System Risk Management," November 18, 2010; http://web.mit.edu/~alo/www/Papers/ntsb17.pdf

2.   *Ibid*, p 21

3.   National Transportation Safety Board, "The Investigative Process"; https://www.ntsb.gov/investigations/process/Pages/default.aspx

4.   *Ibid*

5.   Without comment on its relevance, it is worth noting that the BOP final Accident Investigation Report was issued only four days after the BOP stack was pulled from the water.

6.   Associated Press, "Gulf oil spill testimony to Congress: Not our fault," May 11, 2010; http://www.nola.com/news/gulf-oil-spill/index.ssf/2010/05/gulf_oil_spill_testimony_to_co.html

7.   C-Span, "Gulf of Mexico Oil Spill, Part 1," May 12, 2010, at time mark 3:14; https://www.c-span.org/video/?293463-1/gulf-mexico-oil-spill-part-1

8.   Philip J. Hilts, "Tobacco Chiefs Say Cigarettes Aren't Addictive, April 15, 1994; http://www.nytimes.com/1994/04/15/us/tobacco-chiefs-say-cigarettes-aren-t-addictive.html?pagewanted=all

9.   John M. Broder, "BP's Chief Offers Answers, but Not to Liking of House Committee," The New York Times, June 17, 2010; http://www.nytimes.com/2010/06/18/us/politics/18spill.html

10.   Passages cited are from Findings of Fact and Conclusions of Law, Phase One Trial, Judge Barbier and Mag. Judge Shushan; https://www.epa.gov/sites/production/files/2014-10/documents/phaseonetrial.pdf

11.   Derek Hawkins, "Macondo Workers Broke with Oilfield Practices, Expert Says," Law360, March 6, 2013; https://www.law360.com/articles/421562/macondo-workers-broke-with-oilfield-practices-expert-says

12.   Findings of Fact, Ibid

13.   United States of America v. BP Exploration & Production, Inc., Guilty Plea Agreement, Exhibit A; https://www.justice.gov/iso/opa/resources/43320121115143613990027.pdf

14.   United States of America v. BP Exploration & Production, Inc., Guilty Plea Agreement, p 5; https://www.justice.gov/iso/opa/resources/43320121115143613990027.pdf

15.   David Uhlmann, "After the Spill is Gone: The Gulf of Mexico, Environmental Crime, and Criminal Law," Michigan Law Review, Vol. 109: 1413, p 1454; https://repository.law.umich.edu/cgi/viewcontent.cgi?referer=https://www.google.com/&httpsredir=1&article=1784&context=articles

16.   Jesse Eisinger, The Chickenshit Club: Why the Justice Department Fails to Prosecute Executives (New York, NY: Simon & Schuster, 2017), p 318.

Chapter Eight

1.   United States District Court Eastern District of Louisiana; United States of America v. Robert Kaluza and Donald Vidrine, "Superseding Indictment for Involuntary Manslaughter, Seaman's Manslaughter and Clean Water Act" (issued November 15, 2012), p 8; https://www.justice.gov/iso/opa/resources/25201211151436387 43323.pdf

2.   18 U.S.C. 1115 – Misconduct or Neglect of Ship Officers, para 1; https://www.gpo.gov/fdsys/pkg/USCODE-2010-title18/pdf/USCODE-2010-title18-partI-chap51-sec1115.pdf

3.   Jeanne M. Grasso, "Law and Order: The Emergence of the Seaman's Manslaughter Statute," Benedict's Maritime Bulletin, Second Quarter 2005; http://38.98.220.44/siteFiles/Publications/490C53F432B10A6F13741CE1A4E43E6C.pdf

4.   Ibid

5.   Kurt Orzeck, "US Can't Revive Seaman's Manslaughter Claims in BP Spill," Law360, March 11, 2015; https://www.law360.com/articles/630509/us-can-t-revive-seaman-s-manslaughter-claims-in-bp-spill

6. US Department of Justice spokesman Peter Carr as quoted by Keith Goldberg, "BP Deepwater Supervisors Beat Manslaughter Rap," *Law360*, December 2, 2015; https://www.law360.com/articles/733426/bp-deepwater-supervisors-beat-manslaughter-rap

7. David Uhlmann as quoted by Janet McConnaughey and Michael Kunzelman, "Manslaughter charges dropped for BP supervisors in oil spill," Associated Press, December 3, 2015

8. United States District Court Eastern District of Louisiana; United States of America (Docket No. 12-CR-265) v. Robert Kaluza; Transcript of Trial Proceedings heard before the Honorable Stanwood R. Duval, Jr.; Proceedings Morning Session, February 17, 2016

9. United States District Court Eastern District of Louisiana; United States of America (Docket No. 12-CR-265) v. Robert Kaluza; Transcript of Trial Proceedings heard before the Honorable Stanwood R. Duval, Jr.; Charles Jones testimony, February 24, 2016, p 1374-1375

10. *Ibid*, p 1375-1376

11. *Ibid*, p 1391

12. *Ibid*

13. Harry R. Weber, "Key oil spill evidence raised to surface," Associated Press article as run by *The Hour*, September 4, 2010; http://www.thehour.com/norwalk/article/Key-oil-spill-evidence-raised-to-surface-8303573.php

14. Jones testimony, *Ibid*, pp 1393-1395

15. *Ibid*, p 1403

16. *Ibid*, p 1410

17. *Ibid*, p 1420

18. *Ibid*, p 1417

19. *Ibid*, p 1442-1443

20. In the United States District Court For the Eastern District of Louisiana, United States of America v. Robert Kaluza, Case No. 2:12-cr-00265, Stipulation No. 3; Document 423, Filed February 22, 2016

21. Robert Hinton as featured by Mark Curriden, *Dallas Morning News*, March 7

Chapter Nine

1. Marilyn Vos Savant, "Parade Columnist Marilyn Vos Savant's Favorites Questions from the Last 30 Years," Parade, December 23, 2016; https://parade.com/533287/marilynvossavant/parade-columnist-marilyn-vos-savants-favorite-questions-from-the-last-30-years/

2. Stephen Dinan, "Holder heading out, but contempt lives on," *The Washington Times*, September 25, 2014 (https://www.washingtontimes.com/news/2014/sep/25/contempt-congress-case-will-proceed-without-holder/

3. "BP Exploration and Production Inc. Agrees to Plead Guilty to Felony Manslaughter, Environmental Crimes and Obstruction of Congress Surrounding Deepwater Horizon Incident: BP Agrees to Pay a Record $4 Billion in Criminal Fines and Penalties Two Highest-Ranking BP Supervisors on Deepwater Horizon Oil Rig Charged with Manslaughter and Former Senior BP Executive Charged with Obstruction of Congress", The United States Department of Justice, November 15, 2012; https://www.justice.gov/opa/pr/bp-exploration-and-production-inc-agrees-plead-guilty-felony-manslaughter-environmental

4. BP Sustainability Review 2009, p 5; https://www.bp.com/content/dam/bp/pdf/sustainability/group-reports/bp_sustainability_review_2009.pdf

5. *Ibid*, p 29

6. *Ibid*, p 29

7. Shané Schutte, "CEOs that used staff as scapegoats to protect the firm – or themselves," *RealBusiness*, March 9, 2016; https://realbusiness.co.uk/hr-and-management/2016/03/09/ceos-that-used-staff-as-scapegoats-to-protect-the-firm-or-themselves/

8. David M. Uhlmann, "After the Spill Is Gone: The Gulf of Mexico, Environmental Crime, and the Criminal Law," June 2011, Abstract; https://deepblue.lib.umich.edu/bitstream/handle/2027.42/107454/109MichLRev.pdf.txt;jsessionid=FEE96701B6A2568C26D93BD8F7D04D87?sequence=3

9. *Ibid*, Conclusion

10. "Frequently Asked Questions: Corporate Cooperation and the Individual Accountability Policy," The United States Department of Justice; https://www.justice.gov/dag/individual-accountability/faq

11. *Ibid*

12. United States of America v. BP Exploration & Production, Inc., Guilty Plea Agreement, Exhibit C; https://www.justice.gov/iso/opa/resources/43320121115143613990027.pdf

13. *Ibid*

Chapter Ten

1. BP Investigation Report 2010, page 22

2. Per Holand and Pal Skalle, Deepwater Kicks and BOP Performance (SINTEF, July 2001).

3. From an email exchange between John Guide, BP, and Paul Johnson@deepwater.com on April 15, 2010.

4. David Uhlmann, "After the Spill is Gone: The Gulf of Mexico, Environmental Crime, and Criminal Law," *Michigan Law Review*, Vol. 109: 1413, p 1454; https://repository.law.umich.edu/cgi/viewcontent.cgi?referer=https://www.google.com/&httpsredir=1&article=1784&context=articles

INDEX

www.ingramcontent.com/pod-product-compliance
Lightning Source LLC
Chambersburg PA
CBHW031945080426
42735CB00007B/270